Fractals for the Classroom: Strategic Activities Volume

Heinz-Otto Peitgen   Evan Maletsky
Hartmut Jürgens   Terry Perciante
Dietmar Saupe   Lee Yunker

Fractals for the Classroom:
Strategic Activities Volume One

Springer-Verlag
New York Berlin Heidelberg London
Paris Tokyo Hong Kong Barcelona

Heinz-Otto Peitgen
Institut für Dynamische Systeme
Universität Bremen
D-2800 Bremen 33
Federal Republic of Germany *and*
Department of Mathematics
University of California
Santa Cruz, CA 95064 USA

Hartmut Jürgens
Institut für Dynamische Systeme
Universität Bremen
D-2800 Bremen 33
Federal Republic of Germany

Dietmar Saupe
Institut für Dynamische Systeme
Universität Bremen
D-2800 Bremen 33
Federal Republic of Germany

Evan Maletsky
Department of Mathematics and
   Computer Science
Montclair State College
Upper Montclair, NJ 07043 USA

Terry Perciante
Department of Mathematics
Wheaton College
Wheaton, IL 60187-5593 USA

Lee Yunker
Department of Mathematics
West Chicago Community High School
West Chicago, IL 60185 USA

TI-81 Graphics Calculator is a product of Texas Instruments Inc.

Casio™ is a registered trademark of Casio Computer Co. Ltd.

Camera-ready copy supplied by the authors.
Printed and bound by John D. Lucas Printing Co., Baltimore, MD.
Printed in the United States of America.

9 8 7 6 5 4 3 2 1

ISBN 0-387-97346-X Springer-Verlag New York Berlin Heidelberg
ISBN 3-540-97346-X Springer-Verlag Berlin Heidelberg New York

Preface

There are many reasons for writing this first volume of strategic activities on fractals. The most pervasive is the compelling desire to provide students of mathematics with a set of accessible, hands-on experiences with fractals and their underlying mathematical principles and characteristics. Another is to show how fractals connect to many different aspects of mathematics and how the study of fractals can bring these ideas together. A third is to share the beauty of their structure and shape both through what the eye sees and what the mind visualizes.

Fractals have captured the attention, enthusiasm, and interest of many people around the world. To the casual observer, their color, beauty, and geometric structure captivates the visual senses like few other things they have ever experienced in mathematics. To the computer scientist, fractals offer a rich environment in which to explore, create, and build a new visual world as an artist creating a new work. To the student, fractals bring mathematics out of past history and into the twenty-first century. To the mathematics teacher, fractals offer a unique, new opportunity to illustrate both the dynamics of mathematics and its many connecting links.

It is our desire to give the reader a broad view of the underlying notions behind fractals, chaos, and dynamics. The approach is through a series of strategic activities that reveal these ideas in a non-threatening fashion. They involve the reader directly in constructing, counting, computing, visualizing, and measuring. We have focused on the large number of mathematical connections that exist between fractals and the contemporary mathematics curricula found in our schools, colleges, and universities. Fractals are rich in many fundamental concepts of mathematics that are accessible and taught to most students.

It is our hope that through these strategic activities, students will find an added enjoyment for mathematics and an added awareness of the world around them. Their experiences with the fascinating and challenging topic of fractals will drive their curiosity, stretch their imagination, and pique their interest.

Through the writing of this first of several volumes of strategic activities on fractals, we have sought to bring together a collaborative international effort between the cutting edge of mathematical research and contemporary mathematics education. We hope this publication brings new material to classroom teaching and new excitement to the learning of mathematics world wide.

Evan M. Maletsky   Heinz-Otto Peitgen
Terry Perciante   Dietmar Saupe
Lee E. Yunker   Hartmut Jürgens
USA, 1991   Germany, 1991

Authors

Hartmut Jürgens. *1955 in Bremen (Germany). Dr. rer. nat 1983 at the University of Bremen. Research in dynamical systems, mathematical computer graphics and experimental mathematics. Employment in the computer industry 1984-85, since 1985 Director of the Dynamical Systems Graphics Laboratory at the University of Bremen. Author and editor of several publications on chaos and fractals.

Evan M. Maletsky. *1932 in Pompton Lakes, New Jersey (USA). Ph. D. in Mathematics Education, New York University 1961. Professor of Mathematics, Montclair State College, Upper Montclair, New Jersey, since 1957. Author, editor, and lecturer on mathematics curriculum and materials for the elementary and secondary school, with special interest in geometry.

Heinz-Otto Peitgen. *1945 in Bruch (Germany). Dr. rer. nat. 1973, Habilitation 1976, both from the University of Bonn. Research on nonlinear analysis and dynamical systems. 1977 Professor of Mathematics at the University of Bremen and since 1985 also Professor of Mathematics at the University of California at Santa Cruz. Visiting Professor in Belgium, Italy, Mexico and USA. Author and editor of several publications on chaos and fractals.

Terence H. Perciante. *1945 in Vancouver (Canada). Ed.D in Mathematics Education, State University of New York at Buffalo, 1972. Professor of Mathematics at Wheaton College, Illinois, since 1972. Received a 1989-90 award for Teaching Excellence and Campus Leadership from the Foundation for Independent Higher Education. Author of several books, study guides and articles at the college level.

Dietmar Saupe. *1954 in Bremen (Germany). Dr. rer. nat 1982 at the University of Bremen. Visiting Assistant Professor of Mathematics at the University of California at Santa Cruz, 1985-87 and since 1987 Assistant Professor at the University of Bremen. Research in dynamical systems, mathematical computer graphics and experimaental mathematics. Author and editor of several publications on chaos and fractals.

Lee E. Yunker. *1941 in Mokena, Illinois (USA). 1963 B. S. in Mathematics, Elmhurst College, Elmhurst, Illinois. 1967 M. Ed. in Mathematics, University of Illinois, Urbana. 1963 - present: Mathematics Teacher and Department Chairman, West Chicago Community High School, West Chicago, Illinois. Member of the Board of Directors of the National Council of Teachers of Mathematics. Presidential Award for Excellence in Mathematics Teaching 1985. Author of several textbooks at the secondary level on geometry and advanced mathematics.

Table of Contents

Preface v
Authors vi
Table of Contents vii
Connections to the Curriculum ix
Foreword by Benoit B. Mandelbrot xi

Unit 1   **SELF-SIMILARITY**
           Key Objectives, Notions, and Connections........................... 1
           Mathematical Background 2
           Using the Activities Sheets 6
           1.1   Sierpinski Triangle and Variations 11
           1.2   Number Patterns and Variations 13
           1.3   Square Gasket 15
           1.4   Sierpinski Tetrahedron 17
           1.5   Trees.. 19
           1.6   Self-Similarity: Basic Properties 21
           1.7   Self-Similarity: Specifics... 23
           1.8   Box Self-Similarity: Grasping the Limit 25
           1.9   Pascal's Triangle 29
           1.10 Sierpinski Triangle Revisited 31
           1.11 New Coloring Rules and Patterns................................ 33
           1.12 Cellular Automata 35

Unit 2   **THE CHAOS GAME**
           Key Objectives, Notions, and Connections 37
           Mathematical Background 38
           Using the Activities Sheets 43
           2.1   The Chaos Game 47
           2.2   Simulating the Chaos Game 49
           2.3   Addresses in Triangles and Trees 51
           2.4   Chaos Game and Sierpinski Triangle 53
           2.5   Chaos Game Analysis 55
           2.6   Sampling and the Chaos Game 57
           2.7   Probability and the Chaos Game 59
           2.8   Trees and the Cantor Set 61
           2.9   Trees and the Sierpinski Triangle 65

Unit 3   **COMPLEXITY**
           Key Objectives, Notions, and Connections 69
           Mathematical Background 70
           Using the Activities Sheets 75
           3.1   Construction and Complexity 79
           3.2   Fractal Curves 81
           3.3   Curve Fitting 83
           3.4   Curve Fitting Using Logs 89
           3.5   Curve Fitting Using Technology.................................. 91
           3.6   Box Dimension 93
           3.7   Box Dimension and Coastlines 99
           3.8   Box Dimension for Self-Similar Objects 101
           3.9   Similarity Dimension 105

Answers 109

Connections to the Curriculum
Unit 1

CONNECTIONS	1,1	1,2	1,3	1,4	1,5	1,6	1,7	1,8	1,9	1,10	1,11	1,12
Congruency		▨	▨	▨		▨						
Ratios				▨	▨							
Numerical patterns		▨	▨		▨							
Geometric patterns	▨	▨	▨	▨	▨	▨	▨	▨	▨	▨	▨	▨
Geometric sequences		▨	▨		▨							
Pascal's triangle									▨	▨	▨	▨
Probability												
Similarity	▨	▨	▨	▨	▨	▨		▨		▨		
Area		▨	▨									
Visualization	▨	▨	▨	▨	▨	▨	▨	▨	▨	▨	▨	▨
Modular arithmetic									▨		▨	▨
Coordinate system										▨		
Logic and truth tables									▨			
Power functions												
Linear functions												
Exponential functions										▨		
Logarithms												
Data plotting												
Slope												
Limit concept		▨	▨	▨	▨	▨	▨	▨				
Box and whisker plots												
Curve fitting												
Goodness of fit												

Students can best see the power and beauty of mathematics when they view it as an integrated whole. This chart shows the many connections these strategic activities have to established topics in the contemporary mathematics program.

The National Council of Teachers of Mathematics, in its *Curriculum and Evaluation Standards for School Mathematics*, stresses the importance of mathematical connections:

> *"The mathematics curriculum should include investigation of the connections and interplay among various mathematical topics and their applications so that all students can -*
>
> - *recognize equivalent representations of the same concept;*
> - *relate procedures in one representation to procedures in an equivalent representation;*
> - *use and value the connections among mathematical topics;*
> - *use and value the connections between mathematics and other disciplines."*

Connections to the Curriculum
Units 2 and 3

CONNECTIONS	2,1	2,2	2,3	2,4	2,5	2,6	2,7	2,8	2,9		3,1	3,2	3,3	3,4	3,5	3,6	3,7	3,8	3,9
Congruency																			
Ratios												X				X	X	X	
Numerical patterns					X	X	X	X				X		X	X	X	X	X	X
Geometric patterns	X	X	X	X			X	X			X	X	X	X		X	X	X	X
Geometric sequences				X			X	X			X					X	X	X	
Pascal's triangle																			
Probability	X	X			X	X	X												
Similarity											X	X							
Area																X	X	X	
Visualization	X	X	X	X			X	X	X		X	X	X	X	X				
Modular arithmetic																			
Coordinate system		X											X	X	X	X	X	X	
Logic and truth tables																			
Power functions													X	X		X	X	X	
Linear functions													X	X	X				
Exponential functions													X	X	X				
Logarithms													X		X	X	X	X	
Data plotting													X	X	X	X	X	X	X
Slope													X	X		X	X	X	X
Limit concept		X	X		X		X	X			X	X							
Box and whisker plots						X													
Curve fitting													X	X	X	X	X	X	
Goodness of fit															X		X	X	

Foreword

When I opened the *New York Times* this morning, my eye jumped to a story that must be shared immediately with the readers of this book. In the Westinghouse Science Talent Search for 1991, the first place was won by a young lady from North Carolina who submitted a paper on fractal geometry! She determined the dimensions of fractals generated by Pascal's triangle and its higher dimension analogs.

Benoit Mandelbrot

Yesterday, her topic was known to a very small number of specialists, and it could almost be called obscure, even though there was one scholarly monograph on the subject. This monograph was published in Tashkent by the Academy of Sciences of Uzbekistan (in the Russian language, not the Uzbek!). Tomorrow, however, the substance of the winning project will be known to every student who is fortunate to use this book. To help students match the determination and hard work of the Westinghouse winner remains, of course, the responsibility of parents and teachers. But my friends who authored this book have done a splendid job; thanks to them, the parents and the teachers now have all the help they may need.

All these developments have come as the sweetest compliment to my ears. No token of esteem from my peers could go faster to my heart. As a result, my mind wanders back to my own school days. They were lived in the shadow of war and catastrophe, but have left me with mostly happy recollections. Yet, I do not recall ever wishing to be a student again. Now this new series of textbooks and workbooks fills me with wistfulness, and I wish to thank and congratulate the authors for their achievement.

I am reminded also that friends tell me of a rumor they hear more and more often: fractal geometry being, all things considered, simple and natural, it must have been around almost forever, like counting, and I must have lived long ago. My friends find they are not always believed when reporting that they know me, that I am not even old, and that I continue happily to ride the very same horse I had conjured in my youth. Thus, I find myself a step ahead of Mark Twain, when he cabled to the Associated Press that "the report of his death was an exaggeration". I chuckle at the thought that this lovely Volume will challenge my friends to even greater efforts to prove my existence.

The theme of when and why also has a serious aspect the educator may find worth considering. The persons who place me far in the dim past implicitly believe that science had a Golden Age before the Fall. In this view, much of today's science seems to be long past the Fall, since the same *New York Times* reports that scientists spend billions of dollars to dig a hole in Texas and to organize huge teams to bring this hole to life. In this same view, the Golden Age was when the world and science were simple and easy. It was enough to lean down and pick treasures from the very surface of a virgin soil, and scientists could give birth to new thoughts without pain.

But the history of thought teaches us that there was no such Golden Age. Every bit of science — even that which looks easy after the fact — has had to be discovered in agony. In particular, before one leans down to pick treasures, one must be able to distinguish them from dross. Mother Nature, to paraphrase Einstein, is refined but not evil, and does seem on occasion to be playful, as when the deepest truths are hidden in the manner of the Edgar Alan Poe story, *The Purloined Letter*: right in front of all eyes. Unfortunately, human babies are unable to see anything, even though — unlike baby cats — they are born with open eyes. To see is a skill that both cats and humans must be taught, and this finally brings me to my last point.

The human activities that have had most to do with the eye are painting and geometry, and each went through several periods that were perceived as a Golden Age; they did not last. Today, thanks to the unforeseeable and uncanny help of computer graphics, geometry is rising again, and I cheer with my whole soul. My good fortune has been, first, that my eye was easy to train, and second, that during the long historical period when teachers were being advised to forget geometry and neglect the eye, various good or otherwise bad events had combined to spare my love for geometry and my dedication to its service. Today, I believe deeply that science could accommodate more variety, and badly needs many more people with a sharp and happy roving eye. I hope that, in the hands of dedicated teachers supplied with good textbooks, fractal geometry will help train those indispensable people. To help achieve this goal is among the highest callings I can imagine.

Benoit B. Mandelbrot
IBM Research
and Yale University
March 5, 1991

Unit 1
Self-Similarity

KEY OBJECTIVES, NOTIONS, and CONNECTIONS

The activities in this unit show a dynamic interplay between numerical patterns and geometric patterns. The specific details introduced in this chapter develop the notion of self-similarity, a feature that is characteristic of many fractals. Some of the outcomes appear to be generated from completely random procedures. Yet, these random processes can produce surprising results in the form of highly structured patterns exhibiting the beautiful aspects of self-similarity.

Connections to the Curriculum

The material covered in these strategic activities form an integral part of a contemporary mathematics program. They may be included in a single unit on the topic or integrated into the existing curriculum through those areas to which they are connected.

PRIMARY CONNECTIONS:

Congruency	Similarity
Ratios	Area
Numerical Patterns	Pascal's Triangle
Geometric Pattern	Limit Concept

SECONDARY CONNECTIONS:

Visualization	Modular Arithmetic
Logic and Truth Tables	

Underlying Notions

Self-Similarity

If parts of a figure contain small replicas of the whole, then the figure is called self-similar. If the figure can be decomposed into parts which are exact replicas of the whole, then the figure is called strictly self-similar. Every part of a strictly self-similar structure contains an exact replica of the whole.

Pascal's Triangle

Pascal's triangle is an array of numbers commonly seen as the coefficients of the binomial expansion $(x+y)^n$ as n increases through the whole numbers.

Cellular Automata

Cellular automata is a system or process in which an element of the next level of the system is derived from specific immediate antecedents in the prior level according to some automatic formation rule.

MATHEMATICAL BACKGROUND

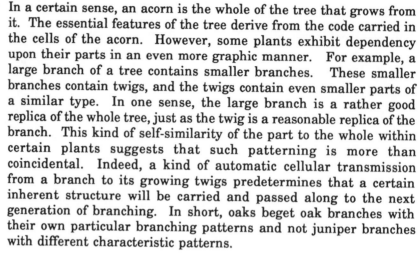

The Bigger Picture

In a certain sense, an acorn is the whole of the tree that grows from it. The essential features of the tree derive from the code carried in the cells of the acorn. However, some plants exhibit dependency upon their parts in an even more graphic manner. For example, a large branch of a tree contains smaller branches. These smaller branches contain twigs, and the twigs contain even smaller parts of a similar type. In one sense, the large branch is a rather good replica of the whole tree, just as the twig is a reasonable replica of the branch. This kind of self-similarity of the part to the whole within certain plants suggests that such patterning is more than coincidental. Indeed, a kind of automatic cellular transmission from a branch to its growing twigs predetermines that a certain inherent structure will be carried and passed along to the next generation of branching. In short, oaks beget oak branches with their own particular branching patterns and not juniper branches with different characteristic patterns.

The underlying idea of self-similarity is immediate and intuitive and can be easily seen and understood from appropriate geometric designs and natural phenomena such as that described for tree branching. However, any rigorous treatment of the topic reveals a close link to the limit concept and requires substantial mathematical knowledge and insight.

Stage 0

Stage 1

Stage 2

Self-similarity: An Intuitive Approach

One approach to understanding the notion of self-similarity involves the ability to apply an iterative rule in a geometric setting. Suppose you start with a rule that supplies a method for repeatedly partitioning a polygon into selected smaller and smaller similar polygons. At each stage in the process, the rule is applied to every remaining polygon. The successive stages continuously generate smaller and smaller similar polygons as the process is repeated over and over, without end. Our interest focuses on the properties present in the figure that would appear after infinitely many iterations.

Consider an equilateral triangle for the initial polygon.
Let the iterative rule be to
   a) connect the midpoints of the sides with line segments, and
   b) remove the middle triangle of the four triangles formed.

At the first stage, three equilateral triangles replace the initial one. At the second stage, the rule is applied to each of these three triangles, subdividing each one into three smaller similar triangles.

Remember, the iterative process requires that the rule be applied repeatedly on all remaining equilateral triangles at each and every stage, once the middle triangles are removed.

Continuing the iterative process in this way leads to a very porous, open figure much like a triangular sieve. In fact, the Activity Sheets show that, as the number of levels in the iterative process increases, the area remaining from the original triangle approaches zero. The final result, generated by infinitely many iterations of the sort described, is called a Sierpinski triangle, named after the great Polish mathematician Waclaw Sierpinski who discussed this object around 1917.

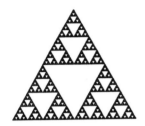

The Sierpinski Triangle

A striking feature of the Sierpinski triangle is its strict self-similarity. The completed Sierpinski triangle naturally decomposes into three triangular parts, an upper portion and two lower portions. Each of these parts is an exact replica of the whole original figure. Likewise, each of these parts itself can be decomposed into three smaller ones, again exact replicas of the original figure. When a figure exhibits this part-to-whole replication, we say that the figure displays self-similarity. Since each and every part of the Sierpinski triangle has this property, it is also strictly self-similar.

Box Self-Similarity

In order to give an operational meaning to the property of self-similarity, we are necessarily restricted to dealing with finite approximations of the limit figure. This is done using the method which we will call box self-similarity where measurements are made on finite stages of the figure using grids of various sizes.

The basic idea is that, at any given finite stage in the generation of a self-similar figure, a point is reached beyond which examinations with grids of finer resolutions give box counts that do not detect any additional differences in the figure.

As another view of this property, imagine placing a transparent grid of squares or pixels over a given stage of the figure. Shade in and count each pixel that contain any part of the figure. On a separate grid, shade in and count those pixels that contain any part of a suitably enlarged portion of the figure. Repeat the process with successive stages of the figure. For the final version of the figure to be self-similar, you must be able to reach a stage where these two pixel counts and pixel patterns become identical irrespective of the size of the selected grid.

Pascal's Triangle

Pascal's triangle is commonly seen as a triangular array of numerical coefficients in the binomial expansion $(x+y)^n$ where the exponent increases through the whole numbers from 0 to n. This triangular array of numbers offers a rich setting for studying both numerical and geometric patterns.

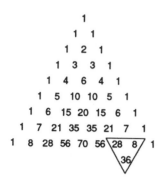

A simple iterative algorithm for generating an entry in the numerical array of numbers in Pascal's triangle is to add the two numbers in the level above it. For example, the number 36, shown in the bottom row, is the sum of the two numbers 28 and 8 immediately above it. Unfortunately, the numbers imbedded deeply within the triangle are very large, and this ultimately makes the numerical iteration process increasingly laborious.

We can introduce a coloring procedure into Pascal's triangle that does not depend upon the magnitude of the numbers but only upon knowing which entries in the table are even and which are odd.

A Coloring Rule

First, simplify the notation by writing 0's for table entries that are even and 1's for those that are odd. Generating new entries at the next level only requires our knowing how even and odd numbers combine under addition.

even + even = even	$0+0=0$	mod 2
even + odd = odd	$0+1=1$	mod 2
odd + even = odd	$1+0=1$	mod 2
odd + odd = even	$1+1=0$	mod 2

Modular Arithmetic

The addition of numbers in terms of 0's and 1's as shown above is known as modulo 2 arithmetic. Modulo 2 addition consists of dividing the sum of two numbers by 2 and recording the remainder as either 0 or 1.

The coloring rule is very easy to apply. Simply color each cell in the table white if it contains a 0 and black if it contains a 1. To extend the pattern to a new cell in a new row, look at the coloring of the two cells immediately above it and apply the correct rule from the four listed. A simplified generalization reads:

If the two cells immediately above are colored the same, color the cell white. If the two cells are different, color the cell black.

Sierpinski Revisited

Continuing the coloring pattern leads to another surprise. Unexpectedly, a Sierpinski-like array of black dots emerge in the triangle. As more and more lines are added to Pascal's triangle, the Sierpinski pattern is imbedded deeper and deeper into the resulting figure. Details of the process can be found in the Activity Sheets.

A Coloring Shortcut

Even generating the colors for the cells in Pascal's triangle by an iterative row by row fashion for more than just a few levels is really impractical. However, by using an addressing system, similar to addresses in a grid of city streets, we can find which color belongs to any given cell in any given row immediately. Each cell in the grid can be described by a two-coordinate number pair.

(left coordinate , right coordinate)

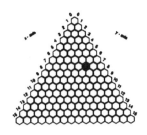

Using the axes shown, the shaded cell has the coordinate address (2,5).

Now convert each coordinate into a binary number. The coordinate address (2,5) becomes (010,101) in binary form. Write the binary coordinates of that cell, one directly above the other. If any column has two 1's, then that cell is colored white. Otherwise the cell is colored black. The cell marked in the example is colored black because no column of corresponding binary digits contains two 1's.

$$010$$
$$101$$

Logic

Comparing binary digits in this fashion is actually equivalent to performing a logical *and* where both of two conditions must be true for the conclusion to be true.

T and T -> T	1 and 1 -> 1
T and F -> F	1 and 0 -> 0
F and T -> F	0 and 1 -> 0
F and F -> F	0 and 0 -> 0

Cellular Automata

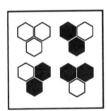

We have seen from the Pascal's triangle illustration that coloring rules can be formulated numerically. Another approach is found in computer graphics applications and very recent numerical simulation methods. Cellular automata utilize coloring look-up tables that visually specify how a cell in the next level should be colored based upon those antecedents in the prior level that are related to it. The Activity Sheets include a cellular automata coloring look-up table that generates the same Sierpinski-like triangle seen when coloring Pascal's triangle.

Additional Readings

Fractals for the Classroom, Chapter 2

USING THE ACTIVITY SHEETS   UNIT 1

1.1 Sierpinski Triangle and Variation

Specific Directions   The iterative algorithm in Activity 1.1A calls for connecting the midpoints of the sides of each triangle and keeping only the three resulting smaller triangles in the corners. Count dots to locate the midpoints.   As the process is repeated, remember to apply it to all triangles that appear.   Geometrically, from one triangle, you generate 3, 9, 27, and then 81 smaller and smaller triangles over the first four stages. Encourage careful counting of the dots on the grids in each activity.   Activity 1.1B generates a variation of the Sierpinski triangle.

Implicit Discoveries At any given stage in these activities, the new triangles produced are all congruent.   As the number of steps in the iterative process increases, the areas remaining from the original figures decrease towards zero.   Each activity has an algorithm that, if repeated without end, leads to a deterministic fractal.   In Activity 1.1A, the figure approaches the Sierpinski triangle, a particularly intriguing fractal that is the basis for all the activities of this chapter.

The vertices of the triangles generated at any stage in these activities identify points that remain parts of the final fractal.   It is essential to view these iterative geometric processes as continuing on without end and to visualize the resulting figures.

1.2 Number Patterns

Specific Directions Use the figures given to check constructions from the last activity and develop a visual feeling for the geometric changes made in going through successive levels.   The corresponding numerical changes serve as the focus of this activity.

Use the drawings to generate number patterns that can be extended and generalized.

Implicit Discoveries   Tie the two sets of table entries on each sheet to geometric sequences.   In Activity 1.2A, the number of triangles is repeatedly multiplied by 3 in going from one stage to the next.   The corresponding geometric sequence has a ratio of 3 for successive terms.   Since $r > 1$, the sequence diverges.   On the other hand, to find successive areas, repeatedly multiply by 3/4.   This geometric sequence has a ratio of 3/4 for successive terms.   Since $r < 1$, the sequence converges to 0.

Another observation, especially important to the development of this entire chapter, is the repeated application of an iterative process without end.   You cannot get the Sierpinski triangle or the final square gasket by stopping at any particular level.   The analogy is the difference between an infinite geometric sequence and some finite portion, starting from the beginning.   Their sums are essentially never the same.

Extensions   Complete the number and area tables for the triangle again using an algorithm that uses only the center triangle at each stage and not the corner ones.   Here the geometry and the algebra are more comfortably connected visually.

1.3 Square Carpet

Specific Directions   This activity generates a variation of the Sierpinski triangle. The square dot grid produces a geometric figure based on squares.   At each level, only the eight squares around the boundary are used for the next level.   The iterative process starts with one square and generates

8, 64, and then 512 smaller and smaller squares. Encourage careful counting of the dots on the grids in each activity.

Implicit Discoveries   At any given stage in these activities, the new squares produced are all congruent. As the number of steps in the iterative process increases, the areas remaining from the original figures decrease towards zero. As in Activity 1.1 this algorithm, if repeated without end, leads to a deterministic fractal.

Extensions   1. Mentally extend the iterative process on the square gasket and conjecture on the nature of the resulting shape of the next stage, level 4, and the number of holes it will contain.
2. Study the number patterns developed for the square carpet if the iterative process uses only the corner squares at each successive stage.

1.4 Sierpinski Tetrahedron

Specific Directions   Three-dimensional visualization is a very important skill being developed in this activity. Different people will see different things in the initial figure shown for Activity 1.4.

Be sure to accurately count the dots needed to draw the midpoint segments in the next stage of the figure. It is only necessary to connect those midpoints on visible faces. The resulting figure is viewed as a pyramid made from a stack of small tetrahedrons, 9 on the bottom, two layers of 3 each in the middle, and then 1 on the top. Shading in all the right faces of the tetrahedrons one way and all the left faces another way helps make the figure appear as a three-dimensional object.

Implicit Discoveries   The activity involves understanding an iterative process that is geometric. At each level, a tetrahedron is replaced by four smaller ones, half the size. The remaining space, an octahedron, is never used at any stage. Inherent in this repeating geometric process is a geometric sequence with ratio, $r = 4$, that counts tetrahedrons at each stage.

$$1 \quad 4 \quad 16 \quad 64 \quad 256 \quad \quad \ldots \quad \quad 4^n \quad \ldots$$

As the iterative process is repeated over and over, the number of tetrahedrons gets larger and larger without end while the space they occupy gets smaller and smaller without end. Only at the limit does the complete Sierpinski tetrahedron appear. It is literally full of holes, a three dimensional version of the two-dimensional Sierpinski triangle.

Extensions   1. Verbally describe the shape and name the figure removed at each stage in the iterative process.
2. Explore related number patterns such as the total volumes at successive stages. In the limit, the Sierpinski tetrahedron occupies three-dimensional space, with nonzero length, width, and height, but its volume is zero.

1.5 Trees

Specific Directions   Be sure branches are drawn in the correct length and direction at each stage. The triangular dot paper facilitates construction. New growth at each stage goes off 60° left and right from that of the last stage. The angles between every pair of segments from a common endpoint is 120°.

Observe that, however long the growth continues, the tree will never extend beyond the hexagonal boundary.

Implicit Discoveries   The complete tree, in its limiting state, has some very intriguing visual properties. Each segment can be viewed as the trunk of its own tree, couched in its own smaller hexagonal boundary.   Look for these successively smaller but complete trees enclosed in successively smaller hexagons.   All these trees are exact images of the initial tree.   All trees of all sizes in the figure have the same number of branches.

Selected paths through the complete tree also have intriguing visual properties.   In particular, the spirals, always turning clockwise or counterclockwise, have lengths tied to geometric series. Their numbers are also tied to geometric series.   Two spirals start at the base of the trunk.   Four start at the first branching point, eight from the second branching points, and so on.

1.6 and 1.7   Self-Similarity

Specific Directions   Consider the cover of a book that contains a picture of a hand holding that very book. The cover of the book shown in the picture must contain the very same hand holding the book which on its own cover shows yet another hand holding the book and so on. As we look deeper and deeper into the scene, we will always see yet another cover that will be exactly like every other cover before or after it.   This is the property of self-similarity.

A similar situation arises when you hold a mirror facing another mirror and look at the scene being reflected back and forth.   This self-similarity is a property of all fractals.   It appears in much of nature as well.   The small top of a stalk of cauliflower or broccoli looks very much like the top of the entire head or bunch.   If readily available, you might want to obtain mirrors or broccoli in order to actually experiment with self-similarity.

Implicit Discoveries   Just as the complete Sierpinski triangle occurs only in the final stage, so too for the tree.   That is to say, while each branch appears to look like both smaller and larger branches, they are, if fact, all exact replicas of the whole only at the final stage when the process is carried out *ad infinitum*.   Since that stage is well beyond the precision possible with pencil and paper construction, the exactness required for self-similarity in this activity must ultimately be visualized mentally.

As the tree grows through more and more stages, its height and width are restricted within tight limits while the total length of its branches becomes large without bound.

Extensions   Tie the geometric iteration of tree growth to geometric sequences. Assume the initial vertical "trunk" of the level 0 tree is 4 inches.

> 1. At what level will the lengths of the shortest branches be less than 1/1000 of an inch?
> 2. What is the maximum height of the tree, growing in this very special way?

1.8 Box Self-Similarity

Specific Directions   On Activity 1.8A, shade those figures in the left hand column as instructed. Then do the same for the middle column of figures, and finally for the last column. On Activity 1.8B, shade the grids for both stage 1 figures before proceeding in order to stage 2, 3, and 4. Do the same for Activity 1.8C and D.

Implicit Discoveries   Two tests using box counting can be utilized to check a figure for strict self-similarity.   The first test applies increasingly refined grids to increasingly refined levels of the figure.   If each stage is a replica of the part of the next stage, then box counting, as illustrated in Activity 1.8A, will determine this by identifying a stage at which a given grid size will fail to detect differences in the successive underlying figures.   This will occur irrespective of the selected grid size.

1.9 through 1.11   Pascal's Triangle and the Sierpinski Triangle

Specific Directions   Review the iterative process of generating numbers in Pascal's triangle. While the algorithm works at every level, the numbers quickly grow large.   You only need to know if an entry is odd or even to find the connection between Pascal's triangle and Sierpinski's triangle in these activities.

Emphasize the parallelism between addition of evens and odds and modular 2 addition with 0's and 1's.   The transition from numbers to positions is made through coloring.
> white for the 0's   (even entries)
> black for the 1's   (odd entries)

Explore how to extend the coloring pattern from row to row without referring to 0's and 1's or E's and O's.   Try to discover the coloring rule and use it to extend the second array in Activity 1.9B.
> *If the two cells immediately above have the same color,*
> *then color the cell white.   If different, color the cell black.*

Implicit Discoveries   Since Pascal's triangle has this geometric relationship to the Sierpinski triangle, the odd and even properties of its entries must have a similar iterative pattern as the presence or absence of subtriangles in the Sierpinski triangle.   This pattern can be applied from row to row by a simple coloring rule, thus connecting the Pascal array to the coloring tables of cellular automata, covered in the next activity.

Extensions   Count the number of even and odd entrys in each row of the Pascal triangle. Conjecture on a rule that computes these counts. The rule is as follows. The number of odd entries in row   n   is always a power of two. The number of factors of 2 in this power is equal to the number of 1's in the binary expansion of the row number   n. For example, in row 6 there are 4 odd entries, because there are two 1's in the binary expansion 110 of the row number 6. In row 13 there are 8 odd entries; the binary expansion of 13 is 1101. Make a plot showing the counts versus the number of the row and discuss the scaling properties of this graph.

1.12   Cellular Automata

Specific Directions   Review the Pascal relationship where each successive entry is found from the values of the two entries immediately above it. Cellular automata follow similar rules where cells in arrays are colored by rules established according to the coloring of specific cells above them.

The color of each cell in successive rows is found by applying the correct coloring choice from the coloring look-up table.   Be sure to recognize the connection between the completed colored array and those from the Pascal arrays.

Implicit Discoveries   These tables reflect properties seen earlier.   Some patterns that can be generated by this method appear chaotic while others appear controlled.   The process is sensitive to only minor changes in initial conditions.

Note that the underlying coloring process shown requires only binary choices and hence is well suited for current technology.   Cellular automata are widely used in computer applications and modeling.

Extensions   Make your own choices on the coloring of cells in the initial row and the left and right hand boundary cells.   Create your own coloring look-up table and see what occurs.   Compare results, explore, and conjecture on what conditions make the patterns appear regular or random.

1.1 SIERPINSKI TRIANGLE AND VARIATIONS

This construction process, when repeated over and over, generates a well known fractal image called the Sierpinski triangle (or gasket).

Construction   Connect midpoints on the sides as shown, keeping only the three corner subtriangles formed.

Apply the construction process on the newly formed corner subtriangles through a second, third, and fourth stage. Remember, at every stage, each remaining triangle is transformed into three new subtriangles with sides half as long. Three times as many triangles appear at each successive stage. Count dots carefully. Every vertex of every subtriangle at each of the first four stages is on a dot on the grid paper. The resulting figure should contain 81 small triangles, representing the fourth stage in the construction of the *Sierpinski triangle.* Shade in these triangles.

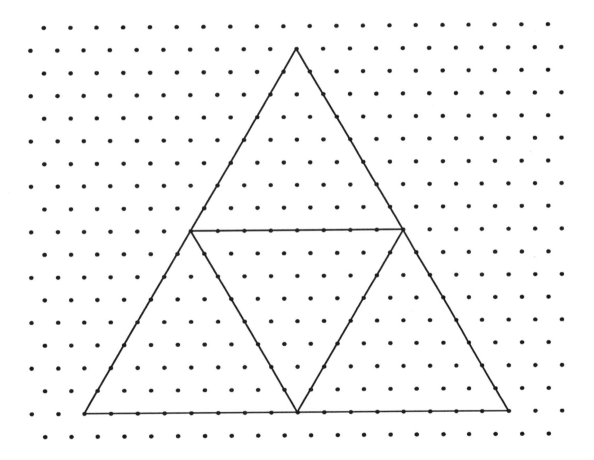

1. Imagine repeating the process.   Visualize and describe how the figure changes. If the process continues on without end, a Sierpinski triangle emerges.

2. What would remain of the original large triangle after four iterations if the algorithm were changed to keeping only the inner triangle?

TRIANGLE VARIATION   1.1B

When repeated over and over, this construct variation generates another fractal.

Construction   Connect trisection points on the sides as shown, keeping
                        only the six border subtriangles.

In this variation, the sides of the triangle are divided into thirds.   Repeat the process
through a second iteration using exactly the same procedure in each of the six border
subtriangles shown in this first stage.   Count dots carefully.   Each vertex of each of the
36 congruent subtriangles that emerge at the second stage are on dots of the grid
paper.   Shade in these triangles.

1. Imagine repeating the process over and over.   At each stage, each triangle is
    transformed into six new subtriangles with sides one-third as long.   Describe what
    you would see of the original triangle if the process were continued on without end.

2. Change the algorithm from keeping the six border subtriangles to keeping the
    three inner ones.   What kind of figure would emerge after two iterations?

1.2   NUMBER PATTERNS WITH VARIATIONS   1.2A

This activity explores some of the number patterns found in the Sierpinski triangle.

DIRECTIONS   The first four stages of the construction of the Sierpinski triangle are shown below.   In subsequent stages, the subdivision continues into smaller and smaller triangles.   Use these figures to explore number patterns that emerge as the Sierpinski triangle is developed through successive iterations.

Stage 0

Stage 1

Stage 2

Stage 3

Stage 4

NUMBER OF TRIANGLES

1. Count the number of shaded triangles at each stage 0 through 4.

STAGE	0	1	2	3	4	5	...	n
NUMBER	1							

2. Extend the pattern to predict the number of triangles at stage 5.
   What constant multiplier can be used to go from one stage to the next?

3. Generalize to find the number of triangles for level n.
   As n becomes large without bound, what happens to the number of triangles?

AREA OF TRIANGLES

4. Let the area at stage 0 be 1.   Find the total shaded areas at stages 1 through 4.

STAGE	0	1	2	3	4	5	...	n
AREA	1							

5. Extend the pattern to predict the total area at stage 5.
   What constant multiplier can be used to go from one stage to the next?

6. Generalize to find the total area at stage n.
   As n becomes large without bound, what happens to the shaded area?

NUMBER PATTERNS FROM TRISECTION

1.2B

Varying the construction algorithm varies not only the figures but the number patterns produced as well.

DIRECTIONS   The first three stages of the triangle construction using the trisection algorithm are shown below.   In subsequent levels, smaller and smaller subtriangles are formed.   Use these figures to explore the number patterns that emerge as more and more iterations are performed on the figure.

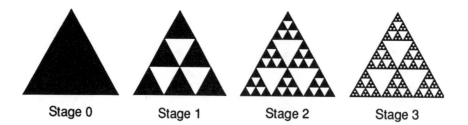

Stage 0   Stage 1   Stage 2   Stage 3

NUMBER OF TRIANGLES

1. Count the number of shaded triangles at each stage 0 through 3.

STAGE	0	1	2	3	4	5	. . .	n
NUMBER	1							

2. Extend the pattern to predict the number of shaded triangles at stages 4 and 5. What constant multiplier can be used to go from one stage to the next?   As n becomes large without bound, what happens to the number of shaded triangles?

3. Compare this number pattern to that for the number of shaded triangles for stages in the Sierpinski triangle.   In which case are the numbers increasing more rapidly?

AREA OF TRIANGLES

4. Let the area at stage 0 be 1.   Find the total shaded area at stages 1 through 3.

STAGE	0	1	2	3	4	5	. . .	n
AREA	1							

5. Extend the pattern to predict the total area at stages 4 and 5.   What constant multiplier can be used to go from one stage to the next?   As n becomes large without bound, what happens to the shaded area?

6. Compare this number pattern to that for the areas for stages in the Sierpinski triangle.   In which case are the areas decreasing more rapidly?

SQUARE CARPET   1.3A

Repeating this construction over and over generates a fractal based on a square.

Construction   Connect trisection points on the sides as shown, keeping
                        only the eight boundary subsquares.

Repeat the process through three successive iterations, the first being shown below.
Remember, use only the border subsquares at each stage.   The result will be a
square carpet made from 512 small subsquares.   Shade in these stage-3 squares.
There should be 73 square holes of three different sizes in the carpet.

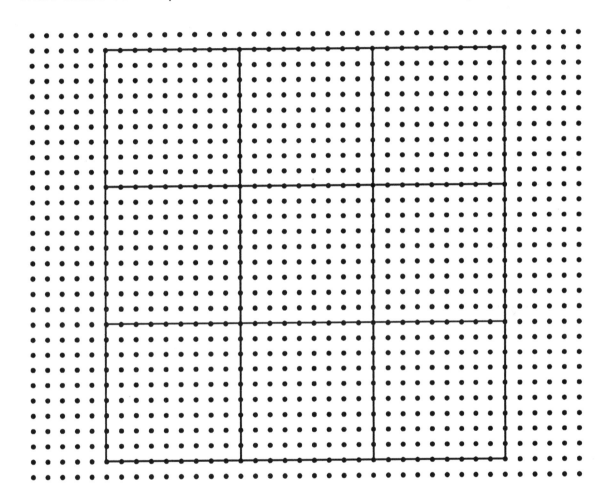

1. Imagine repeating the process over and over.   At every stage, each square is
   transformed into eight new subsquares with sides one-third as long.   Describe
   what you would see of the carpet if the process were to continue without end.   How
   many holes will there be?   How much of the original square will remain?

2. Suppose the algorithm were changed from keeping the eight border subsquares to
   keeping only the four corner subsquares. What figure emerges after two iterations?

NUMBER PATTERNS IN THE SQUARE CARPET   1.3B

This activity focuses on number patterns found in generating successive stages leading to the square carpet.

DIRECTIONS   The first three stages of the construction of the square carpet are shown below.   In subsequent stages, the subdivision continues into smaller and smaller subsquares.   Use these figures to explore number patterns that emerge as the square carpet is developed.

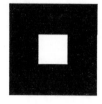

| Stage 0 | Stage 1 | Stage 2 | Stage 3 |

NUMBER OF SQUARES

1. Count the number of shaded subsquares at each stage 0 through 3.

STAGE	0	1	2	3	4	...	n
NUMBER	1						

2. Extend the pattern to predict the number of shaded subsquares at level 4.   What constant multiplier can be used to go from one stage to the next?

3. Generalize to find the number of shaded subsquares for level n.   As n becomes large without bound, what happens to the number of shaded subsquares?

AREA OF SQUARES

4. Let the area at stage 0 be 1.   Find the total shaded area at stages 1 through 3.

STAGE	0	1	2	3	4	...	n
AREA	1						

5. Extend the pattern to predict the shaded area at stage 4.   What constant multiplier can be used to go from one stage to the next?

6. Generalize to find the shaded area for stage n.   As n becomes large without bound, what happens to the shaded area?

7. Imagine successive figures generated by using only the four corner squares at each level.   What would be the new answers for questions 1 through 6?

1.4   SIERPINSKI TETRAHEDRON   1.4A

This activity applies a construction algorithm to a three-dimensional triangular pyramid or tetrahedron.   The resulting visualization experience leads to an interesting fractal.

Construction   Connect the midpoints on the edges of a tetrahedron as shown, keeping the four new smaller tetrahedrons formed at the vertices and removing the remainder of the solid.

Study the figure to be sure you see the four tetrahedrons formed.   Remember, only the visible faces are shown.   Apply the algorithm a second time to each of the four tetrahedrons.   Count dots carefully since each new vertex for the smaller pyramids lies on a dot of the grid.   The 16 resulting pyramids should all appear to have the same size and shape.   Check to see that their corresponding edges are parallel. Then shade in all left-hand faces one way and all right-hand faces another way.

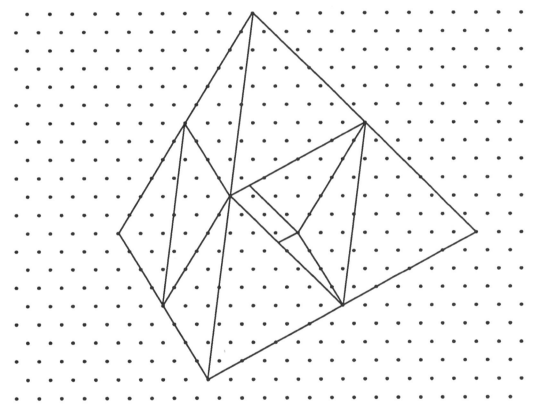

Each tetrahedron has four faces, six edges, and four vertices.

1. In all, how many faces, edges and vertices exist in the 16 resulting stage-2 tetrahedrons of the object, counting shared vertices only once? How many of the faces are visible?

2. How many tetrahedrons would appear at stage 3?   at stage 4?   at stage *n* ? Describe how the figure would change if the process were repeated without end.

3. Use the dot grid to complete the view of one face of a stage-4 Sierpinski tetrahedron.   What figure do you get?

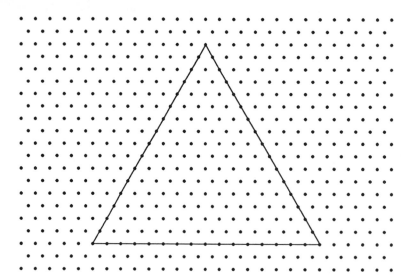

4. A given stage of the Sierpinski tetrahedron takes on difference appearances when viewed from different positions.   Use the dot grid to complete this top view of the stage-1 Sierpinski tetrahedron.   Draw in all visible edges.

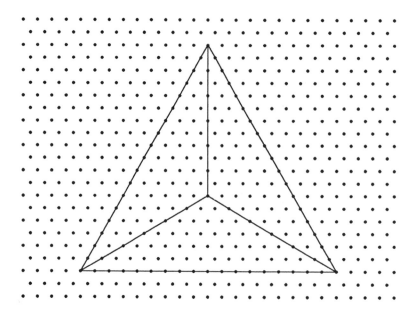

1.5   TREES   1.5A

As trees grow, they branch out. From big branches grow smaller ones.   From these grow smaller ones still, and so on.   Use this dot paper to draw a mathematical tree with some of the same properties as the live ones.

Construction   From the endpoint of each branch, draw two new branches half as long growing off at 60° in opposite direction.

1. Stage 1 of the tree has already been drawn.   Draw the four new branches for stage 2 by connecting endpoints to the appropriate dots on the grid.   Draw the eight new branches for stage 3.   Repeat again for stage 4.   Endpoints should still be on the dots of the grid.   Continue the growing process until the branches become too small to draw.

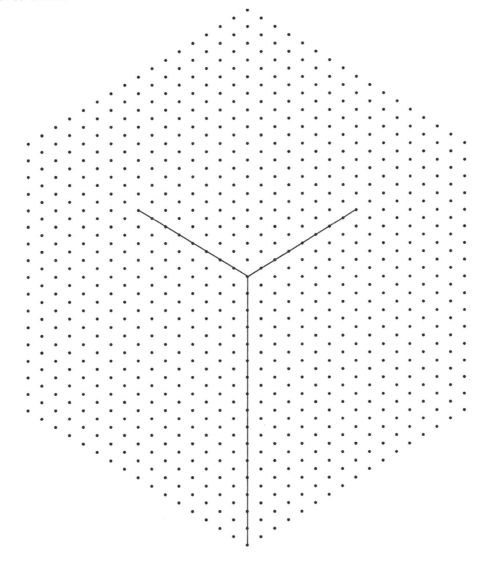

1.5B

Suppose the tree starts with an initial vertical segment of 1 unit as the trunk. Imagine further that the tree continues growing branches, over and over by the process given, until fully grown. Visualize this completed tree.

2. How many branches have lengths of 1/4? of 1/16? What is the sum of the lengths of all branches 1/4 long? 1/16 long?

3. What is the total length of all branches of the completed tree?

4. Are there parts of the completed tree that look like the entire tree? Using the tree just drawn as a model of a fully grown tree, draw a hexagon around a part that would be an exact image of the tree itself. Draw another using a hexagon of a different size.

One interesting shape found on the completed tree is a *spiral*. Start at the base of the tree and turn right at each and every junction point. Note how these particular branches trace out a spiral.

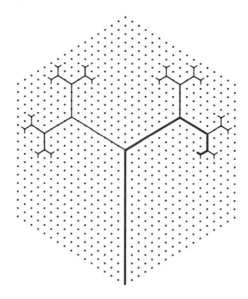

5. Find another spiral that is a reflection of the one just described. What is the length of this spiral?

6. Find four spirals with half the length of the one just described. How many spirals in the tree have one-quarter the length of the original?

7. Consider all such spirals of all sizes that can be found on the tree. They all do not have the same length. Are they all exact replicas of each other except for size?

1.6   SELF-SIMILARITY: BASIC PROPERTIES   1.6A

The Sierpinski triangle displays many of the central features of deterministic fractals. This activity focuses on the property of self-similarity.

> If parts of a figure are small replicas of the whole, then the figure is called *self-similar*.

Some familiar objects in the real world around us have appearances that visually seem to convey this characteristic where parts are small copies of the whole. We want to include these in our intuitive notions of self-similarity. However, there are different degrees of self-similarity.   The highest degree occurs when, within any selected part of a self-similar figure, there exists a replica of the whole figure.

> A figure is *strictly self-similar* if the figure can be decomposed into parts which are exact replicas of the whole. Any arbitrary part contains an exact replica of the whole figure.

1. How would you compare one stalk with an entire head of cauliflower?   In what ways are they alike?   Are they essentially self-similar?   Can you identify another vegetable with similar properties?

Imagine a cover of a book that contains on it a picture of a hand holding that very book. Surprisingly, this somewhat innocent sounding description leads to a cover with a rather complex design.   As we look deeper and deeper into the design, we see more and more of the rectangular covers.

2. What can you say about the number of rectangles representing covers that would appear on the rectangular cover of the actual book?

3. Will every rectangle on the cover have another contained within it?   Will every rectangle, except the cover itself, be contained within a rectangle?

4. In each of the following figures, a part of the completed book cover has been identified. In which cases do the identified parts contain replicas of the whole? Does the completed book cover exhibit either self-similarity or strict self-similarity?

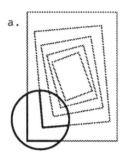 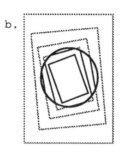 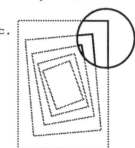

Imagine a tree that grows by spreading out in similar sets of branches as shown. At each stage of growth, the tree looks much the same as at other stages. As more and more stages of growth appear, they look more and more alike. Our eyes begin to detect the essential visual characteristics of self-similarity.

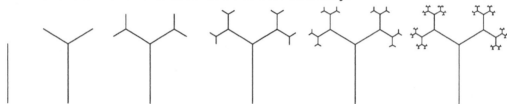

5. Imagine the tree at some finite stage of growth, not yet fully grown. Is there any part of the tree that is an exact replica of the entire tree at that stage? This means, for example, that the part must have the same number of branches as the whole. Can the tree be self-similar at that finite stage?

6. Something very different happens when the tree is complete and fully grown. It is self-similar because it now contains parts that are exact, small copies of the whole. But does every part of the tree contain a copy of the tree? Is the tree strictly self-similar? On the final diagram above, draw a circle around a portion of the tree that, even on the completed version, will fail to contain a replica of the whole tree.

7. Consider the set of leaves at the endpoints of the fully grown tree. Is the set of points self-similar? Is it strictly self-similar?

As a final example, consider the Sierpinski triangle.

8. If you took any part of the Sierpinski triangle, any piece at all of any size that contains some shaded part, would it necessarily contain a replica of the entire triangle? Is the Sierpinski triangle strictly self-similar?

1.7   SELF-SIMILARITY: SPECIFICS   1.7A

What exactly is meant by *a part is an exact replica of the whole* ? This activity
expands on the question and supplies an answer.

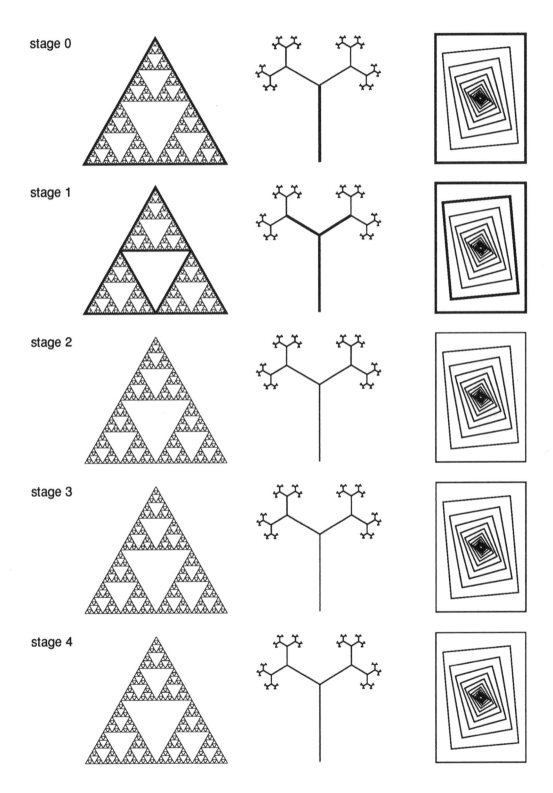

stage 0

stage 1

stage 2

stage 3

stage 4

1. Consider the iterative processes of drawing successive stages in the cover design, the growing tree, and the developing Sierpinski triangle.   Stages 0 and 1 are already highlighted.   Do the same for the parts in stages 2, 3, and 4.

2. As more and more stages are drawn, where are the new parts appearing in the cover design?   in the tree?   in the Sierpinski triangle?

All three figures, in their completed states, are self-similar.   However, the rectangles are converging to a point.   Only parts containing that point will contain *a replica of the whole*.   The trees are branching out toward their final leaves.   Only parts containing those leaves will contain *a replica of the whole*.   But the Sierpinski triangle is simply growing everywhere.   Hence, any of its parts, in whatever size and location you choose, must contain a replica of the whole.   It is this special property that makes the Sierpinski triangle *strictly self-similar.*

1.8   BOX SELF-SIMILARITY: GRASPING THE LIMIT   1.8A

The method of box self-similarity uses finite approximations of the limit object in various grid resolutions to test for self-similarity.

1. For each construction stage of the Sierpinski triangle, shade in fully each square in the grid that contains any of its parts.   This amounts to shading in or lighting the pixels in the grid that intersect the underlying figure.

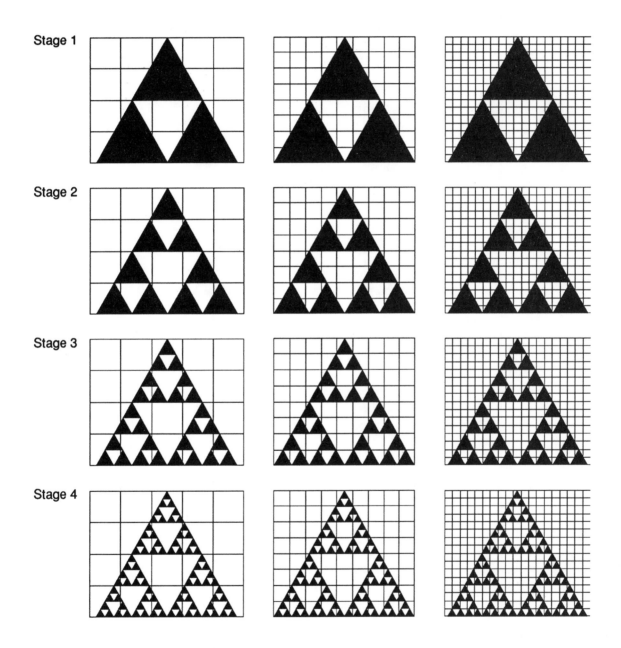

2. At what stage of the figure does the coarse grid stop detecting any difference in the figures?   When does this happen for the medium grid?   for the fine grid?

Here the screen has a coarse resolution.   **1.8B**

3. Each pair of figures shows a stage of the Sierpinski triangle and the lower left part of it, enlarged by a factor of two. Shade the square pixels that cover some portion of the underlying figure. Enter the total number of pixels in the table for the whole figure and then for the corresponding section.

4. In the table, the number of pixels lit for the whole and the section are the same beginning at what stage? What would you expect to happen with a finer grid?

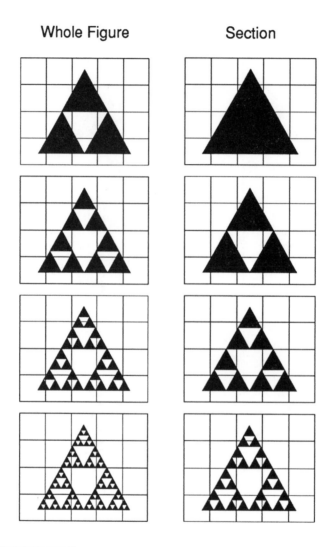

Whole Figure   Section

Stage number	Number of pixels lit		Are the identical pixels lit?
	whole	section	
1			
2			
3			
4	12	12	yes

Here the screen has a medium resolution.   **1.8C**

5. Each pair of figures shows a stage of the Sierpinski triangle and an enlarged section of it. Shade the square pixels that cover some portion of the underlying figure. Enter the total number of pixels in the table for the whole figure and then for the corresponding section.

6. In the table, the number of pixels lit for the whole and the section are the same beginning at what stage? What would you expect to happen with a finer grid?

Whole Figure   Section

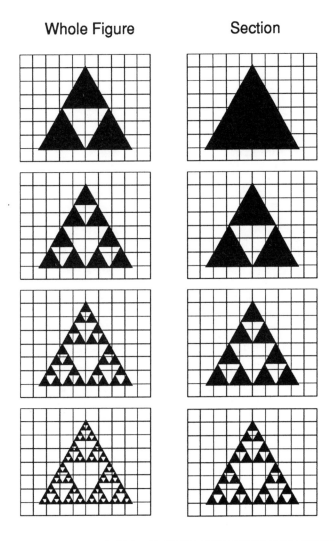

Stage number	Number of pixels lit		Are the identical pixels lit?
	whole	section	
1			
2			
3			
4	36	36	yes

Here the screen has a fine resolution.   **1.8D**

7. Each pair of figures shows a stage of the Sierpinski triangle and an enlarged section of it.   Shade the square pixels that cover some portion of the underlying figure.   Enter the total number of pixels in the table for the whole figure and then for the corresponding section.

8. In the table, the number of pixels lit for the whole and the section are the same beginning at what stage?   Does it appear that, however fine the grid, there will always be some stage beyond which the two counts will be the same?   When this is the case, and the two pixel patterns are identical, then the figure is self-similar.

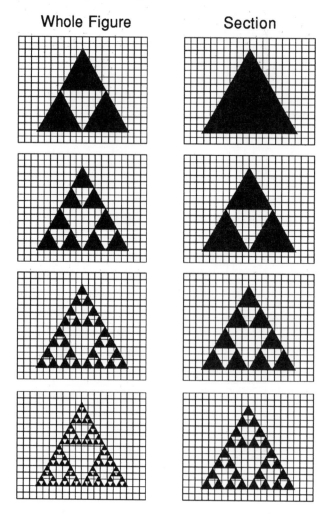

Stage number	Number of pixels lit		Are the identical pixels lit?
	whole	section	
1			
2			
3			
4	102	102	yes

1.9   PASCAL'S TRIANGLE   1.9A

This activity centers around the famous array of numbers called Pascal's triangle. These numbers have been used to solve various probability problems.   The connection here is to the Sierpinski triangle and fractals.

The first number in the initial row 0 of Pascal's triangle is 1.   Every number thereafter is the sum of the two numbers immediately above it.   If only one number occurs in the preceding row, assume the other to be 0.   The triangle is completed through row 10.

1. How many numbers are in row 8?   Row 9?   Row 10? How many will be in row n ?

2. Enter the numbers needed for rows 11 and 12.   Can the numbers be extended? Can the numbers in row n be used to generate those in row $n + 1$?

3. Start with the 1 in row 0 and imagine a vertical line down through the array.   Look at the numbers on opposite sides of the line in each row.   What do you observe?

4. In rows 13, 14, and 15, enter only the letters E for *even* or O for *odd* .   Do not compute the numerical values but rather use these relationships:

$$E + E = E \qquad E + O = O \qquad O + E = O \qquad O + O = E$$

Row

0   1

1   1   1

2   1   2   1

3   1   3   3   1

4   1   4   6   4   1

5   1   5   10   10   5   1

6   1   6   15   20   15   6   1

7   1   7   21   35   35   21   7   1

8   1   8   28   56   70   56   28   8   1

9   1   9   36   84   126   126   84   36   9   1

10   1   10   45   120   210 252   210 120   45   10   1

11   __ __ __ __ __ __ __ __ __ __ __ __

12   __ __ __ __ __ __ __ __ __ __ __ __ __

13   ....

14   ....

15   ....

1.9B

The addition of even and odd numbers leads to modulo 2 arithmetic.   In modulo 2 arithmetic, only the remainders after division by 2 are relevant.   For example, consider $5 + 7 = 12$ and $5 + 8 = 13$.   The sum 12 has a 0 remainder and the sum 13 has a 1 remainder modulo 2.

$$5 + 7 = 0 \mod 2 \qquad 5 + 8 = 1 \mod 2$$

Since the only possible remainders on division by 2 are 0 and 1, every sum modulo 2 must be either 0 or 1.   This is equivalent to saying every sum must be even or odd.

$E + E = E$	$0 + 0 = 0 \mod 2$
$E + O = O$	$0 + 1 = 1 \mod 2$
$O + E = O$	$1 + 0 = 1 \mod 2$
$O + O = E$	$1 + 1 = 0 \mod 2$

5. Enter 0 or 1 in the first eight rows of Pascal's triangle by writing a 0 if the table entry is even and 1 if it is odd.

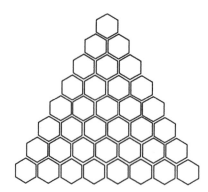

6. This time color in the first eight rows of Pascal's triangle by shading in the entries with 1's (odds) and leaving unshaded the entries with 0's (evens).

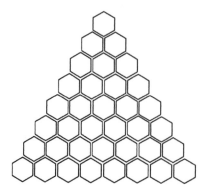

7. Additional rows in the triangle can be colored by the same process using 0's and 1's or evens and odds.   Study the coloring in the triangle above and give a rule for coloring each cell based upon the coloring of the two cells immediately above it.

1.10   SIERPINSKI TRIANGLE REVISITED

The rule for coloring the cells in Pascal's triangle can be stated this way:
> If the two cells directly above are different in color, then shade in the cell so the color is black.   If they are the same in color, leave the cell unshaded so the color is white.   End cells in each row are always colored black.

The first eight rows of the triangular array below have been colored using this rule.

1. Do you see a geometric pattern in the first four rows of the display?   How is it related to stage 1 in the construction of the Sierpinski triangle?

2. How are the first eight rows related to the first four rows?   How are they related to stage 2 of the Sierpinski triangle?

3. Follow the rule above and color in the next eight rows on the triangle.   What stage of the Sierpinski triangle appears from the completed figure?

4. How many rows would be needed in all to represent stage 4 of the Sierpinski triangle?   stage 5?

STAGE	1	2	3	4	5	. . .	n
NUMBER OF ROWS	2	4	8				

5. Now generalize the results in the table.   How many rows are needed for stage n ?

A COLORING SHORTCUT   1.10B

The question arises as to whether or not there is a direct way of finding the coloring of any given cell in Pascal's triangle without running the process through all rows above it. The answers is yes, but the process requires a binary coding of each location.

Start with the origin (0 , 0) as the top entry in the triangular array. Let the x -axis be diagonal to the left and the y -axis diagonal to the right. Then each pair of coordinates (x , y) corresponds to a specific location in the array. Cell A has coordinates (2,1).

1. Give the coordinates for cells B, C, and D.

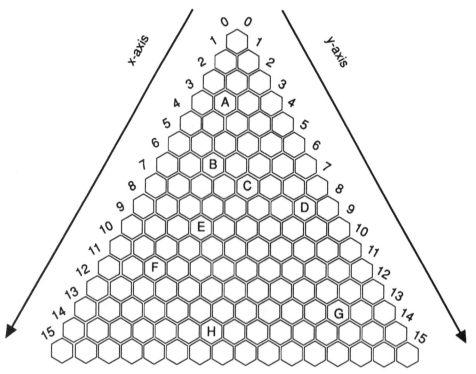

To determine the color of a given cell, place the binary expansions of the two coordinates of the cell above each other and follow this rule:
        If two 1's appear above each other in any one of the columns,
        then the cell is left white. Otherwise, it is shaded in as black.

2. The 4-digit binary coordinates for cell E are (0110,0011). When placed above each other, do any columns have two 1's? Will the cell be colored black or white?

3. What is the color of cell F?   of cell G?   of cell H?

Convert these coordinates to binary form. Then determine if the corresponding cells are colored black or white.

4. (7 , 9)   5.   (12 , 16)   6.   (25 , 40)

1.11 NEW COLORING RULES AND PATTERNS

In this enrichment activity, a modified coloring system is applied to entries in Pascal's triangle based on modulo 3 arithmetic.   A new, but predictable, pattern emerges in the coloring of the cells.

In modulo 3 arithmetic, only the remainders after division by 3 are of interest.   As an example, consider $5 + 7 = 12$, $5 + 8 = 13$, and $5 + 9 = 14$.   The sum 12 has a 0 remainder, the sum 13 has a 1 remainder, and the sum 14 has a 2 remainder.

$$5 + 7 = 0 \mod 3 \qquad 5 + 8 = 1 \mod 3 \qquad 5 + 9 = 2 \mod 3$$

Since the only possible remainders upon division by 3 are 0, 1, and 2, *every* sum modulo 3 must be 0, 1, or 2.

1. Refer to the numerical entries in rows 0 through 8 of Pascal's triangle.   Express each number in modulo 3 form and then color in the corresponding cell using the following rule:
    >   If the entry is 1 or 2, shade the cell black.   If the entry is 0, leave the cell unshaded as white.

2. Study the coloring on the cells thus far completed.   How does it compare with stage 1 of the Sierpinski triangle variation on Activity sheet 1.1B?

3. Try coloring in the remaining rows by replicating the pattern that you see in the first nine rows.   The pattern that emerges should contain the 18 small triangles found in stage 2 of the Sierpinski triangle variation mentioned above.

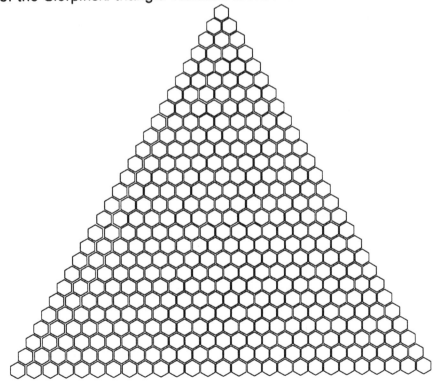

In modulo 9 arithmetic, only the remainders after division by 9 are of interest.

$$5 + 11 = 7 \mod 9 \qquad 5 + 12 = 8 \mod 9 \qquad 5 + 13 = 0 \mod 9$$

Since the only possible remainders upon division by 9 are 0, 1, 2, 3, 4, 5, 6, 7, and 8, every sum modulo 9 must be one of these numbers.   This next activity requires finding the numbers in Pascal's triangle that are divisible by 9 with remainder 0.   These are the numbers equal to 0 mod 9.

4. Refer to the numbers in Pascal's triangle and their mod 9 form.   Color the corresponding cells in this array using the following rule:
   If the entry mod 9 is 0, shade the cell black.
   Otherwise, leave the cell unshaded as white.

See how quickly you can see a coloring pattern emerge that you can follow to complete the array.

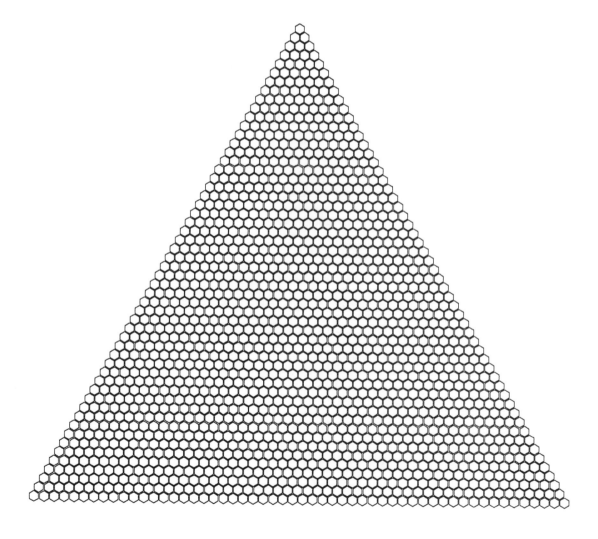

1.12   CELLULAR AUTOMATA

Pascal's triangle represents an evolution process where row after row of cells grow according to specific rules.   Consider coloring the cells by a special rule.   For example, color cells containing odd numbers black and cells containing even numbers white.   Geometric patterns or structures may emerge.   When the rules are changed, new structures are revealed.   Processes of this type are called *cellular automata* .   Numerous relevant applications of cellular automata have been found, including the simulation of fluid flow around obstacles.

This activity is designed to introduce and explore coloring look-up tables.

> A coloring look-up table supplies a visual definition of all the rules needed to color any particular cell based upon the colors of the cells in the row or rows immediately above it.

1. This coloring look-up table has four entries.   State verbally how it tells the coloring of a cell based on the coloring of the two cells immediately above it.

2. Use the rule from the coloring look-up table to complete the coloring of all remaining rows in this array.

3. What coloring pattern do you see?   How is it related to Pascal's triangle?   to the Sierpinski triangle?

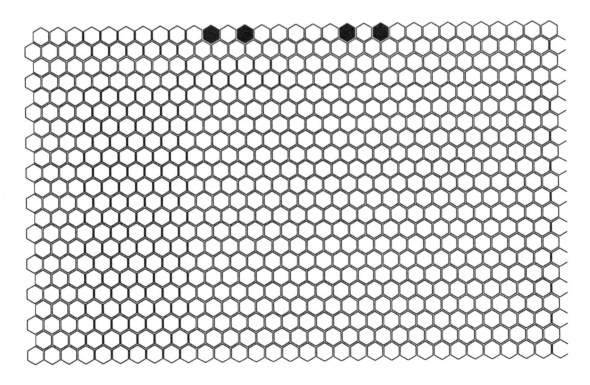

1.12B

Study the coloring of the two successive rows shown from a cellular array. Construct the appropriate look-up table if coloring is based on the two cells immediately above.

4.

5.

6. How many different coloring look-up tables are possible based on two cells? If the coloring of a cell is based on that of the four cells immediately above it, how many different possibilities must be shown in each look-up table?

7. Use the look-up table given to complete the coloring of this cellular array. Where have you seen the resulting structure before?

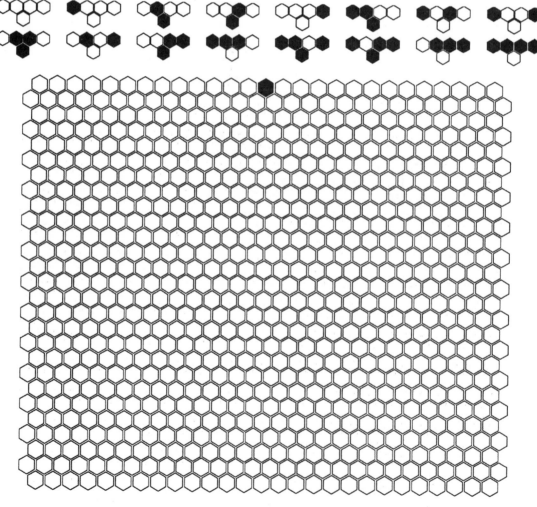

Unit 2
The Chaos Game

KEY OBJECTIVES, NOTIONS, and CONNECTIONS

The activities in this unit connect the chaos game to the Sierpinski triangle. Trees are used to tie addresses for finite sequences of plays in the chaos game to addresses for locations of subtriangles in related stages of the Sierpinski triangle. Infinite strings are eventually considered, and the Cantor set is related to fully grown trees.

The surprising result that the randomly generated chaos game leads to the structured pattern of the Sierpinski triangle illustrates the powerful connections that underly much of mathematics. The study of fractals reveals this beautiful interplay among several mathematical ideas.

Connections to the Curriculum

These materials cover many topics found in contemporary mathematics programs. The activities may be presented separately or integrated into the existing curriculum through those areas to which they are connected.

PRIMARY CONNECTIONS:

Numerical Patterns   Geometric Patterns
Probability   Sampling
Finite and Infinite Sets
Geometric Sequences and Series
Limit Concept

SECONDARY CONNECTIONS:

Visualization   Technology
Binary Numeration   Real Numbers

Underlying Notions

Trees

If points are joined by arcs or line segments, then the resulting network is called a tree if there are no closed regions.

Cantor Set

The Cantor set is the collection of points that remain after removing the middle third of an initial unit segment and then repeatedly removing the middle third of all resulting segments.

Chaos Game

In the random process of the chaos game a point is moved within a triangle leaving a trace of dots, which eventually condenses to a deterministic fractal, the Sierpinski triangle. In each move one of the three vertices of the triangle is chosen at random, and the point moves halfway to the chosen vertex.

Address

By an address consisting of a finite string of symbols subtriangles of the Sierpinski triangle of a given stage are identified. Infinite extensions of such strings yield addresses of points of the fractal.

MATHEMATICAL BACKGROUND

The Bigger Picture

Successive moves in the chaos game are governed by the random rolls of the die. From the unpredictable placement of midpoints within the triangle, a structured pattern emerges that, in the limiting case, becomes the Sierpinski triangle.

Living as we do, in finite surroundings, our senses often control our thoughts. In this unit, the initial addressing is of finite moves in a chaos game, finite paths through the branches of a tree not yet fully grown, and subtriangles in finite stages of a Sierpinski triangle. We have a strong, intuitive feeling for these ideas. We can draw and see them on paper. But the underlying messages of connection come from their limiting, infinite states. These can be theoretically defined but only imagined and visualized in an approximate form in our minds. It is there, in this body of theory and in our conceptual approximations, that fully grown trees exist and the Sierpinski triangle and Cantor set reside. It is the home of many fractal images.

The Chaos Game

Label the three vertices of a triangle L, T, and R for left, top, and right. Let a random roll of a die determine the direction, L, T, or R, for each move. Start at any point inside the triangle. Move halfway to the vertex identified by the roll. From there, move in a similar fashion for subsequent rolls, marking the stopping points as play progresses. This process is called the chaos game.

Repeating this random process of the chaos game *ad infinitum* produces a surprising result, the Sierpinski triangle. This highly structured fractal is generated by infinitely many random rolls of a die!

Addressing Game Moves

Assign L to the faces 1 and 2, T to the faces 3 and 4, and R to the faces 5 and 6 of a die. Start from an arbitrary initial point x_0 and roll the die. Plot the point halfway between x_0 and the vertex named by the die. Call this new point x_1. Roll again and plot the next point

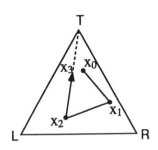

halfway between x_1 and the vertex indicated by this roll. Mark each new midpoint as the game is played.

Now, suppose upon playing the chaos game for three consecutive rolls, an R, L, and T appear in that order. Let $T(L(R(x_0)))$ denote the last point, x_3, in the sequence plotted after starting at the initial point x_0. Specifically, a point x_1 is plotted halfway between x_0 and R, a point x_2 halfway between x_1 and L, and a point x_3 halfway between x_2 and T.

We can simplify the notation by deleting all parentheses. However, when playing the chaos game, understand that the sequence of outcomes TLR is read from right to left and apply to x_0 in this right to left order. Address strings such as TTRLRT, TR, LTTR, and RRL, read from right to left, represent points generated by the chaos game startet with the initial point x_0.

Addressing Triangles

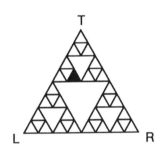

A series of moves in the chaos game can be represented by a sequence of the letters T, L, and R. When read from right to left, each such sequence indicates a path to a point within a particular triangle at the appropriate stage of the Sierpinski triangle. We next look at how we can address these triangles by reading the address strings in the reverse direction.

The triangle address TLR locates a stage-3 triangle that is within the top section T, on the lower left side L, and in the lower right section R of that lower left side.

Triangle addresses can use many letters, specifying triangles very deep into the Sierpinski construction. No matter how long, a triangle address is always read from left to right as an itinerary for a route to the addressed triangle.

Tree Diagram Addresses

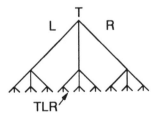

Addresses of triangles in the Sierpinski triangle can be illustrated within the context of tree diagrams. Let TLR identify a path downward through the branches of a level-3 tree. T implies moving from the top straight down to the first corner, L suggests continuing down the left branch to the next corner, and R indicates continuing down the right branch.

The address of each of the 27 stage-3 triangles corresponds to one of the 27 branches of the level-3 tree. The mapping from subtriangles in the Sierpinski triangle to branches in the tree applies at any stage and level. Continuing the branching pattern *ad infinitum* leads to a tree that is itself a fractal, just as the Sierpinski triangle is a fractal.

Comparing Addresses

A direct connection exists between this triangle addressing system and the chaos game. The triangle TLR locates the position of the point when R is rolled first, followed by L and then T. Every triangle in stage 2 of the Sierpinski triangle defines a possible triangular neighborhood of the point after 2 rolls of the die in the chaos game. Every stage-3 triangle defines a possible triangular neighborhood after 3 rolls, and so on. The number of letters in the address that identified the triangular region that contains the point after a given number of plays of the chaos game corresponds to the number of rolls of the die required to reach that subtriangle.

While they are written the same, there is an important difference between the addresses of subtriangles in the Sierpinski triangle and the addresses for points generated by the chaos game. The addresses of triangles are read from *left to right* while points in the chaos game are read from *right to left*. Either interpretation of a string of letters T, L, and R leads to the same subtriangle in the Sierpinski triangle. If the string is interpreted as an address, then it should be read from left to right and the location rule for addresses applied. If the string is interpreted in terms of the chaos game, then it should be read from right to left where the halfway rule applies.

The Connection

As the die is rolled repeatedly without end in the chaos game, you would expect every possible finite sequence of letters to eventually appear. But all of these sequences with any given number of letters, for example n letters, locate all the different triangles in that stage of the Sierpinski triangle (the n-th stage). Thus, letter sequences in the chaos game correspond to subtriangles in the Sierpinski triangle. The longer the game is played, the greater the number of points that appear representing these subtriangles. Played forever, the full Sierpinski triangle appears.

Details of this sampling process and the corresponding argument can be found in the Activity Sheets. Of course, in actual practice, the process can only be applied at an arbitrary finite level.

Probability

TRRL

In the chaos game, the three outcomes T, R, and L have equal, positive probabilities based on the roll of the die. Consider a block of letters formed from a given number of rolls of the die, as for example TRRL. It too must have a positive probability of occurring, whether representing a triangle address or a chaos game address.

The probability associated with the four rolls addressed as TRRL in the chaos game is

$$(1/3)^4 = 1/81$$

This is also the probability of reaching the stage-4 Sierpinski triangle addressed TRRL in any given number of moves of 4 or more.

Connections to the Cantor Set

Begin with a segment described by the set of points in the closed interval [0,1]. Remove, as an open interval, the middle third of the segment. Two segments now remain, each one-third the original length. Using this operation as the iterative process, repeat it over and over again. At each stage, remove the middle third of every segment that remains. The number of segments doubles and their lengths decrease to one-third at each successive stage. The Cantor set is that dusting of points that remains from the original segment when the iterative process is continued *ad infinitum*. It is a deterministic fractal with the property of strict self-similarity.

The distribution of points in the Cantor set corresponds to the leaves on the final boundary of a special two-branch tree that grows *ad infinitum* according to the following rules:

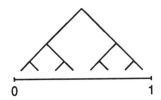

0   1

- At each vertex of the tree, fix the angle measure between both branches at 90°, running 45° left and right off the vertical.

- As you proceed down the tree from level to level, each new branch is 1/3 the length of the previous one.

- To make the final horizontal span of the tree to be 1, the total height must be 1/2. This requires that the vertical height of stage 1 be 1/3.

Geometric Series

$$H = 1/3 + 1/9 + 1/27 + 1/81 + \ldots = 1/2$$

To show that the cardinality of the Cantor set is the same as the real numbers in the unit interval [0,1], a modification of the tree is made. Change the growth factor to 1/2 and the initial stage-1 height to 1/4. Then this tree has the same branching pattern as the one noted above, its leaves can be matched directly to the binary decimals. This establishes the cardinality of the Cantor set to be that of the continuum, as detailed in Activity 2.8.

The last activity uses a three-branch tree with growth factor of 1/4 and an initial, stage-1 height of 3/8. When fully grown, the leaves of this tree form another fractal, much like both the original Cantor set and the Sierpinski triangle.

The fractal just described can be constructed geometrically and without appealing to the tree diagram. Partition a line segment into eight equal sections and remove the third and sixth sections. Each remaining segment, in turn, has the third and sixth of its eight equal subsections removed. Repeat the subdivision and removal process on the remaining segments. The first three stages are illustrated here.

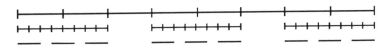

The process is like a one-dimensional version of that used to generate the Sierpinski triangle. Here, each successive stage is formed by subdividing and removing specific parts of existing line segments. In the Sierpinski triangle, each successive stage is formed by subdividing and removing the center parts of existing triangles.

Unfortunately, there is not a one-to-one correspondence between the final leaves of this tree and the points in the Sierpinski triangle. Two choices exist for an exact mapping.

• Modify the Sierpinski triangle using a trisection process. Here, the triangles generated are always separated. This occurs when the half-way rule of the chaos game is changed to a two-thirds rule.

• Modify the tree construction into a different form.

Details of both options can be found in Activity 2.9.

Note that not all modifications of the chaos game produce a fractal shape. For example, choose the four vertices of a square in place of the vertices L, T, and R of the triangle. In each move select one of the four vertices at random and move the game point halfway to that vertex. This procedure will generate random points all over the square. In fact, eventually the complete square will be filled. Of course, the square is not a fractal.

Additional Readings

Fractals for the Classroom, Chapters 5 and 6.

USING THE ACTIVITY SHEETS   **UNIT 2**

2.1 The Chaos Game

Specific Directions The initial activity calls for drawing the successive traces from one midpoint to the next to practice the process. Play the chaos game by marking only the successive, individual midpoints. Choose different starting points so that similarities and differences in the results can be observed. Be sure to measure each successive new midpoint from the last midpoint just located.

For a vivid display of pooled results, plot different sets of points on separate sheets of acetate, superimpose them, and project the combined results. Well defined clustering will occur eventually, but only for large numbers of repeated plays.

Implicit Discoveries Drawing the traces connecting midpoints shows the sequence of plays and vividly displays the randomness of the process. Plotting only the successive midpoints allows for more rapid play and sets the stage for a discovery coming in the next activity.

2.2 Simulating the Chaos Game

Specific Directions Be careful entering and changing the programming code when using the graphing calculator to simulate the chaos game. Do not stress the programming details. Implementation of a simulation on the machine is what is important here.

Implicit Discoveries When traces from one point to the next are drawn on the graphing calculator, the visual image appears to be random. When endpoints alone are plotted, a Sierpinski triangle approximation emerges quite rapidly. The argument supporting this observation requires an understanding of addressing techniques, both for sequences of plays in the chaos game and for triangle locations in successive stages of the Sierpinski triangle. These are developed in the following activities.

2.3 Addresses in Triangles and Trees

Specific Directions Sheet 2.3A develops an addressing procedure for locating the different subtriangles at any given stage in the Sierpinski triangle. The number of letters in the address matches the specific stage used. Sheet 2.3B develops a similar addressing procedure for the different paths through the branches of the tree diagram. Again, the number of letters in the address matches the specific level of tree used.

Implicit Discoveries One quickly recognizes the connection between the addresses of the subtriangles at a given stage and those of the paths through the branches of a tree at the same level.

When read left to right, a triangle address works deeper and deeper into smaller and smaller triangles of successive stages of the Sierpinski triangle. The corresponding tree address, read left to right, moves down through smaller and smaller branches of the tree.

2.4 Chaos Game and Sierpinski Triangle

Specific Directions Use sheet 2.4A to see graphically how paths are traced out by successive rolls in the chaos game. It shows how the first midpoint to vertex R from any location in the original triangle must fall in that triangle R. The midpoint from any point within triangle R to vertex L must fall in triangle LR.   Again, the next midpoint from triangle LR to vertex T must fall in triangle TLR.   The code does not give the exact location of the point but rather the triangle within whose boundaries the point must lie.   Read the game code from right to left in order to match the location TLR, read left to right, for the corresponding Sierpinski triangle.

Implicit Discoveries   The location of a sequence of points plotted in a chaos game depends basically on the random results of the rolls of the die.   When compared, different game sheets will vary widely from one to another.   Yet this apparently random, chaotic behavior has an underlying fractal pattern that controls it.   Even on a graphing calculator or computer screen, every trace, whether repeated by the same person or someone else, is formed by the random number generator of the machine.   While each one is different, it quickly becomes apparent that a large number of plays of the chaos game zeroes in on a surprising fractal result, the Sierpinski triangle.

Extensions   1. In what fractional part of the area of the original triangle of the chaos game is the fifth midpoint restricted?
          2. Discuss the address and successive midpoint locations in the chaos game for the particular sequence of rolls, 3,3,3,3,3.

2.5   Chaos Game Analysis

Specific Directions   Use Activity 2.5 to focus on the number of subtriangles at any stage of the Sierpinski triangle and the equal number of corresponding addresses.   At stage 2 there are nine subtriangles and nine 2-letter addresses identifying sequences of rolls.   Clearly, after two rolls, only one such stage-2 triangle is identified.   However, it is important to note that, to reach all these stage-2 subtriangles, the corresponding sequences of two rolls producing the 2-letter addresses need only appear somewhere in an extended string of plays.   In the next activity, sample data are given on the actual number of rolls taken in each of fifteen different games to reach every stage-2 triangle.

Implicit Discoveries   The argument presented discusses reaching every stage-2 triangle in the Sierpinski triangle through successive and extended plays of the chaos game.   Intuitively, this seems evident.   As more and more plays are made, there is an increasing likelihood that each and every needed 2-letter sequence will appear.   Let n be the number of plays and P be the probability that all stage-2 triangles are reached.   Then as $n \rightarrow \infty$, it follows that $P \rightarrow 1$.

Of course, to complete the argument, it is necessary to establish that all subtriangles at all stages be lit.   Indeed, the Sierpinski triangle itself appears only at the infinite stage.   In actual practice, the process at best can only be applied at some small, restricted finite stage.   If you stop after playing any finite number of times, small or large, the Sierpinski triangle will be incomplete.

2.6   Sampling and the Chaos Game

Specific Directions   This activity deals directly with the question of how many plays of the chaos game are needed to reach each and every stage-2 triangle in the Sierpinski triangle. The approach is through sampling.

On sheet 2.6B new sampling results are presented in a box-and-whisker plot.   Be sure that the stage-2 triangles that correspond to the new 2-letter addresses are shaded in on the screen provided as they appear.   This gives a strong visual interpretation of the process.

Implicit Discoveries This activity begins by presenting the actual results of 15 successive chaos games. The number of plays required ranges from a low of 12 to a high of 41, with a median at 22. This expected number of 22 is based solely on the randomly generated results of these 15 games. Computation of the mathematical expectation directly is a far more challenging exercise.

2.7 Probability and the Chaos Game

Probabilities are used to predict the chances of ending in a specific subtriangle on a given number of rolls. Consideration is then given to the changes created by modifying the probability associated with each event L, T, or R in the chaos game. The results of these changes appear vividly on the graphing calculator.

Implicit Discoveries Keep in mind that every point reached in the chaos game belongs, not only in a specific subtriangle at that stage, but at the same time to a subtriangle at each and every stage beyond that point. The analysis of stage-n triangles applies *only* for rolls numbering n or more.

The results of the last question on sheet 2.7B may come as a surprise. Intuitively, it may not be obvious that the same Sierpinski triangle must emerge on the graphing calculator. The only effect a change in probabilities at the vertices has is to alter the speed that the figure appears around these vertices. It does not change the ultimate, resulting figure. The one exception occurs when one or two of the probabilities assigned are 0.

It is important to see and sense the connecting ideas in mathematics. A study of chaos and fractals offers an excellent opportunity to accomplish this. Note the interrelationship of the many ideas presented to this point in this unit .

2.8 Trees and the Cantor Set

Specific Directions This activity and the next offer an optional extension to some of the ideas thus far developed. While the general size and shape of the tree diagrams drawn earlier were not important, they are critical here. The two-branch trees considered here have right and left branches move out in 45° angles from the vertical. The fully grown trees must have vertical heights of 1/2 and horizontal spans of 1.

In analyzing the intervals spanned as successive sets of new branches grow from the tree, one sees geometrically an iterative process in action. At every new level, the middle third of each existing interval is removed. This is the same algorithm that generates, in its final state, the Cantor set.

Implicit Discoveries As successive sets of new branches grow on the first tree on sheet 2.8A, they span smaller and smaller intervals. In its limit state, the tree is a fractal that exhibits self-similarity. The final endpoints or leaves form a dusting of points along a horizontal unit segment 1/2 unit below the top of the tree. They form a strictly self-similar fractal, better known as the Cantor set.

A modified tree with a growth rate of 1/4 generates points that correspond to the binary decimals over the interval [0,1]. Details can be found in the activity sheets. Since there is a one-to-one mapping between the paths in the two trees, the Cantor set must have the cardinality of the interval [0,1].

2.9 Trees and the Sierpinski Triangle

The branches in the first tree grow such that each successive branch has a height 1/4 that of the one before it.   The right and left branches move out in 45° angles but there is also a center branch. Again, the total span is 1 and the total height 1/2.   Relate the center, vertical branch from the top to the horizontal line at the extreme bottom of the completed tree to the notation for the path $T\overline{TTT}$ . . . Then relate its length to the geometric series

$$3/8 + 3/32 + 3/128 + \ldots = 1/2$$

Will any branches meet in this tree?   Looking at the corresponding tree geometrically, the answer is clearly no.   Conjecture as to the nature of the distribution of points along the terminal line segment at the bottom of the tree.

The algorithm that applies here repeatedly partitions each interval into eight equal parts, always eliminating the third and the sixth.   These are the addresses of the paths that locate the corresponding boundary points of the initial eight intervals.

$$
\begin{array}{lll}
LLL\overline{L}\ldots & LTT\overline{T}\ldots & LRR\overline{R}\ldots \\
TLL\overline{L}\ldots & TTT\overline{T}\ldots & TRR\overline{R}\ldots \\
RLL\overline{L}\ldots & RTT\overline{T}\ldots & RRR\overline{R}\ldots
\end{array}
$$

Shift to the Sierpinski triangle and find the corresponding locations of the stage-4 triangles that match the first four letters in each address.   Then look for the triangles that match these complete addresses in their entirety.   As sequences become longer and longer and eventually infinite the corresponding subtriangles get smaller and smaller and eventually correspond to points.

Implicit Discoveries   Note that no finite sequence of plays in the chaos game has an address that corresponds to a complete path through the tree to the unit segment at the bottom.   Only at the infinite state do paths down the branches of the tree reach the line.   In other words, there is a correspondence from an infinite path in the tree to a point in the Sierpinski triangle established by running the chaos game. However, if we pick points in the Sierpinski triangle there may be two different routes to it.   For example, the distinct paths $LR\overline{RR}$... and $RL\overline{LL}$... down the tree lead to distinct points on the unit interval but to the same point in the Sierpinski triangle.   Two options are available to establish an exact one-to-one correspondence.   One is to modify the Sierpinski triangle to separate all subtriangles at every stage.   The other is to modify the tree diagram.   Both options are presented in detail.

2.1   THE CHAOS GAME   2.1A

Play this game and watch the apparent chaotic behavior of a moving point.

Start with any point inside the triangle formed by vertices L, T, and R.

Step 1   Roll the die and move according to these rules.
                    For 1 or 2, move halfway to L.
                    For 3 or 4, move halfway to T.
                    For 5 or 6, move halfway to R.
Step 2   Connect the point to this newly located midpoint.

Step 3   Starting from the last midpoint located, repeat the steps.

Continue repeating the process extending the path to four successive midpoints.

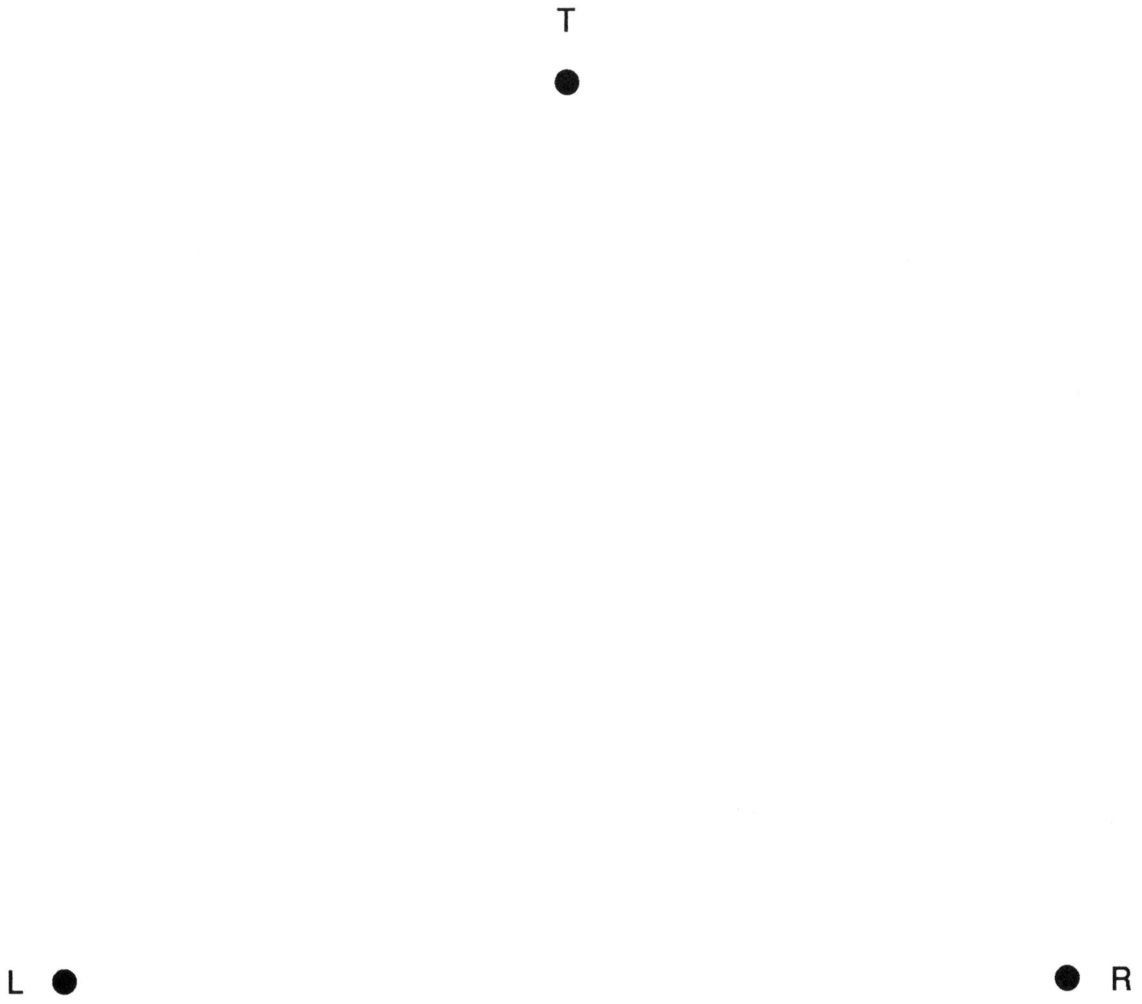

T

●

L ●   ● R

2.1B

Play the chaos game again.   This time mark only the midpoints.   Start with any point inside the triangle formed by points T, L, and R.   Continue repeating the process until at least 20 successive midpoints have been plotted.

For 1 or 2, move halfway to L.
For 3 or 4, move halfway to T.
For 5 or 6, move halfway to R.

T
●

L ●   ● R

1. What is the boundary for all possible points in the chaos game?

2. Do the successive midpoints appear to be randomly located within the boundary?

2.2   SIMULATING THE CHAOS GAME   2.2A

A striking feature of the chaos game is that when the random process is repeated a great many times, the resulting arrangement of points approaches a familiar figure. Unfortunately, the pencil and paper process becomes tedious for large numbers of repetitions.   To use technology to simulate the chaos game, key in the appropriate program into your graphing calculator.

		CASIO	TEXAS INSTRUMENTS
1			:ClrDraw
2			:0->Xmin
3			:1->Xmax
4			:0->Ymin
5			:1->Ymax
6		Range 0,1,1,0,1,1	
7		"X="?->X	:Disp "X="
8			:Input X
9		"Y="?->Y	:Disp "Y="
10			:Input Y
11		0->C	:0->C
12		X->A	:X->A
13		Y->B	:Y->B
14		Plot A,B	
15		Lbl 1	:Lbl 1
16		C+1->C	:C+1->C
17		Plot X,Y	:PT-On(X,Y)
18		Line	:Line (A,B,X,Y)
19		X->A	:X->A
20		Y->B	:Y->B
21		Plot A,B	
22		C>1500=>Goto 4	:If C>1500
23			:End
24		Ran#->N	:Rand->N
25		N<.3333=>Goto 2	:If N<.3333
26			:Goto 2
27		N>.6666=>Goto 3	:If N>.6666
28			:Goto 3
29		.5xX->X	:.5X->X
30		.5xY->Y	:.5Y->Y
31		Goto 1	:Goto 1
32		Lbl 2	:Lbl 2
33		.5x(X+1)->X	:.5(X+1)->X
34		.5xY->Y	:.5Y->Y
35		Goto 1	:Goto 1
36		Lbl 3	:Lbl 3
37		.5x(X+.5)->X	:.5(X+.5)->X
38		.5x(Y+1)->Y	:.5(Y+1)->Y
39		Goto 1	:Goto 1
40		Lbl 4	
41		Plot 5,5	
42		Plot 5,6	
43		Line Δ	

1. Run the program. Notice how it traces out the path of the moving point in the chaos game as it goes from one midpoint to the next. Do the traces appear to cover the entire triangular region?

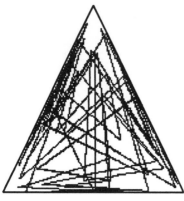 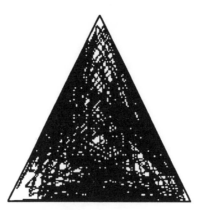 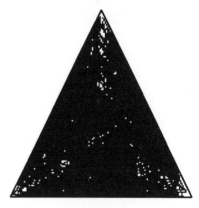

| 100 traces | 500 traces | 1500 traces |

2. Make these changes in the code so that only the positions of the successive points appear on the screen.

    Casio: Delete lines 12, 13, 14, 18, 19, 20, and 21.

    TI-81:   Delete lines 12, 13, 18, 19, and 20.

    Run the program again.   Describe the results.

3. Does it appear that the pattern of points that emerge from the random choices of moves in the chaos game is predictable?

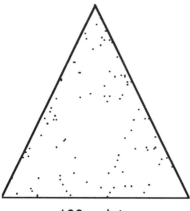 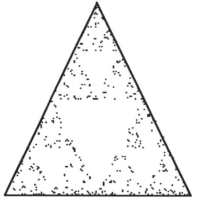 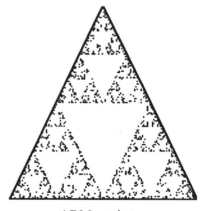

| 100 points | 500 points | 1500 points |

2.3 SIERPINSKI TRIANGLE ADDRESSES

2.3A

Simulating a great many plays of the chaos game on a graphing calculator or computer reveals that the randomly generated sequence of midpoints increasingly produces a highly structured fractal shape. As the random process is repeated without end, the Sierpinski triangle emerges. In order to establish this result, we need a method for addressing specific locations within any stage of the Sierpinski triangle.

Addresses of specific positions or cells within a given stage of the Sierpinski triangle will be expressed using the letters L, T, and R, the vertices of the initial triangle. The number of letters in the address will determine the stage used. For example, the address TLR indicates that the identified triangle is in the top section T (first stage), on the lower left side L (second stage), and then in the lower right section R (third stage) of that lower left side.

TLR

There is a unique location for each address and a unique address for each location.

Shade in the position of the triangle for each address given.

1.

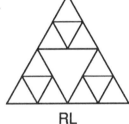

RL

2.

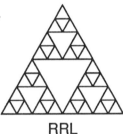

RRL

3.

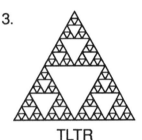

TLTR

Give the address for each triangle marked.

4.

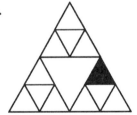

5.

6.

Verbally describe the locations of these stage 4 triangles.

7. TTTT   8. RLTL   9. LRRT   10. TLLL

TREE DIAGRAM ADDRESSES

2.3B

Sequences involving the letters L, T, and R are represented in this activity using tree diagrams.

DIRECTIONS   Consider a tree diagram with three branches emerging from each intersection point.   Every path downward through the tree can be addressed by using a letter L, T, or R for each point.

L means follow the left branch.
T means follow the center branch.
R means follow the right branch.

At level 2, the address LR means move from the top down the left branch and then the right branch.

At level 3, the address TRL means move from the top down the center branch, the right branch, and then the left branch.

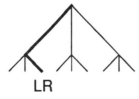

LR

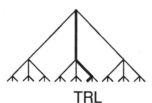

TRL

Give the address for each path shown.

1.

2.
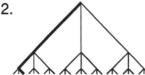

3.

Mark the path for each address given.

4.
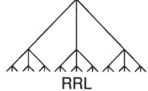
RRL

5.
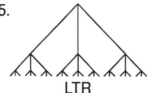
LTR

6.
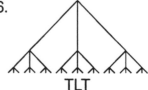
TLT

Arrange these addresses for paths in the order of their final locations, left to right.

7.   TLL   TRL   RTL   LRR   TTT

8.   RTLL   RLTT   TLRT   LRTL   TTLL   LLLL

2.4   CHAOS GAME AND SIERPINSKI TRIANGLE   2.4A

Strings of the letters L, T, and R can represent subtriangle locations at different levels in the Sierpinski triangle and paths in a three-branch tree diagram.   This activity shows how they can also be used to represent successive points generated by the chaos game.

Let the initial point in the chaos game be *some point* x_0 inside the triangle.
If the first roll is R, think of x_1, the halfway point to R, as $R(x_0)$.
The point x_1 must be in the stage-1 triangle R, independent of the location of x_0.

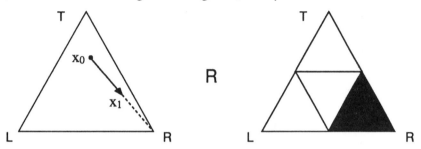

If the second roll is L, think of x_2, the halfway point to L, as $L(R(x_0))$.
The point x_2 must be in the stage-2 triangle LR.

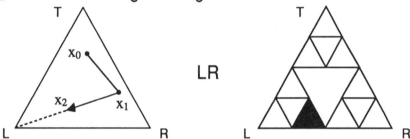

If the third roll is T, think of x_3, the halfway point to T, as $T(L(R(x_0)))$.
The point x_3 must be in the stage-3 triangle TLR.

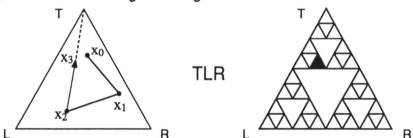

Thus $T(L(R(x_0)))$ denotes the sequence of points, x_1, x_2, x_3.   The notation can be simplified to $TLRx_0$ or simply TLR.   The TLR means R is applied *first* to the initial point x_0, L applied *second* , and T applied *third* .

Caution: How is TLR read?
        When denoting a chaos game sequence of plays, it is read *right to left* .
        When denoting a Sierpinski triangle location, it is read *left to right* .

2.4B

Shade the cell that must contain the point, starting from any point inside the triangle, if

1. the first roll is T.

2. the first roll is L, and the second is T.

3. the first roll is R, and the second and third are L.

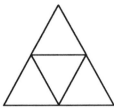

Start with any point in the chaos game triangle.   Shade in the triangle that must contain the point at each stage for the sequence of rolls given.

4. RLTT

 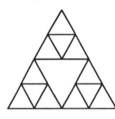  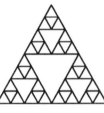

Stage 1   Stage 2   Stage 3   Stage 4

5. TRRL

The site of the point after the third roll in a chaos game is shown.   What was the third roll?   the second roll?   the first roll?

6.

7.

8.

9. Assume that after the second roll, the point in a chaos game is in the cell marked. Show the three possible locations after the third roll.   Show the nine possible locations after the fourth roll.

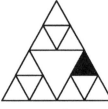
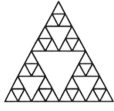
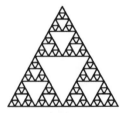

second roll   third roll   fourth roll

2.5   UNDERSTANDING THE CHAOS GAME   2.5A

The Sierpinski triangle exists only in its infinite state.   For all
practical purposes, however, it suffices to visualize it at some
finite stage in its construction.   The chaos game is viewed in
this activity through a finite screen of triangles corresponding
in size and position to those at a given stage in the generation of the Sierpinski triangle.   This is much the same as the
way a television picture is viewed through a screen of dots
called pixels that light up as needed to create the image.

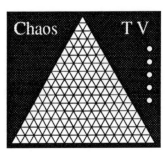

| Stage 0 | Stage 1 | Stage 2 | Stage 3 | Stage 4 |

Stage 1 in the construction of the Sierpinski triangle consists of the three shaded
triangles shown.   To generate this display in the chaos game, view it through a screen
of three triangles of this size in this position.   To light them all, the three outcomes L, T,
and R must occur on rolls of the die.

1. Nine triangles are in the screen for the stage-2 construction of the Sierpinski
   triangle.   Complete the corresponding list of the 2-letter sequences of rolls
   required to light each triangle.
   LL   TL   RL   ___   ___   ___   ___   ___   ___

2. How many triangles are in the screen for the stage-3 construction?   Complete the
   corresponding list of 3-letter strings that show the first roll as a T.   Remember,
   letters in the game code are read right to left.
   LLT   TLT   RLT   ___   ___   ___   ___   ___   ___

3. How many triangles must be lit to show the Sierpinski triangle at the resolution of
   stage 4?   How many of the corresponding 4-letter game code strings will begin at
   the right with R?   with TR?   with TTR?

Consider stage 2 to be the second generation of the Sierpinski triangle.   Look at the
stage 2 triangle shown above.   Its screen consists of 9 small triangles the size of these
second-generation triangles, each to be lit when the chaos game brings the point into
that triangle.   Clearly, a 2-letter game code is needed to represent each second-
generation triangle.   The following investigation will show that for a small triangle to
be reached, its 2-letter code must appear some place in the extended sequence of
iterations or rolls in the chaos game.

Assume the chaos game is played through 10 iterations denoted by this sequence.
LTRRTRLLRT

The first roll is T, and the second, R.   These first two rolls locate a point somewhere in the second-generation triangle with address RT.   The third roll is L.   This moves the point into the triangle with address LR.   The fourth roll is L again, moving the point into the triangle with address LL.   Shifting left in blocks of two letters at a time through the 10-letter string gives the addresses of the successive second-generation triangles reached during these 10 iterations.   Note that two locations, TT and TL, are not yet lit. But they are likely to be reached in successive rolls of the die beyond these ten.

Game code	L T R R T R L L R T	Roll
	R T	2
Addresses	L R	3
of	L L	4
second-	R L	5
generation	T R	6
triangles	R T	7
reached	R R	8
	T R	9
	L T	10

By taking blocks of three letters at a time, the eight addresses of the successive third-generation triangles reached can be readily listed.   Successive blocks of four letters give addresses of fourth-generation triangles reached.   The entire ten-letter string is the address of a single tenth-generation triangle.

4. Give the address of the fourth-generation triangle reached at each successive stage during this sequence of 20 rolls of the die.   List all 17 in order starting with the fourth iteration, LRTT.   How many different addresses of fourth-generation triangles are in the list?

    TRRLRTLRTLLRRRTTLRTT

5. Use this triangular screen to shade in all the fourth-generation triangles that will be lit from the 20 iterations given above.   The first, LRTT, is already entered.

6. If you rolled the die a great many times in the chaos game, producing a very long string of letters L, T, and R, would you expect to cover every fourth-generation triangle in stage 4 of the Sierpinski triangle?   How many triangles are in stage n? If the string were long enough, would you expect to cover all triangles in n-th stage?

2.6   SAMPLING AND THE CHAOS GAME   2.6 A

How many rolls of the die will it take to fall in every subtriangle of the screen for a given stage of the Sierpinski triangle?   This sample required 24 rolls to light each of the nine cells of the Sierpinski triangle at stage 2.

Start here with the first roll . Enter successive rolls to the left.

Each line under the string indicates where a new stage-2 triangle was entered as play progressed.   The first cell lit was RT.   Only after 24 rolls did the point reach the ninth and final cell, LL.

What is the *expected number* of rolls needed to fall in each and every subtriangle in the screen of the Sierpinski triangle at stage 2?   One approach is to conduct many experiments and use the data to predict the answer.   Here are the results of 15 different chaos games played on a stage-2 screen of the Sierpinski triangle.   The numbers recorded indicate how many rolls were needed to reach all nine subtriangles at this level.

    24 33 18 28 13 16 17 41 29 22 12 28 18 13 26

This is a box-and-whisker plot of the same data.   The box locates the middle 50% of the ordered results.   The line in the box locates the median.   The whiskers extend to the extremes.

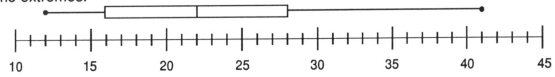

Simulate the game yourself using a die.   Repeatedly roll the die, placing the appropriate entries L, T, and R in the cells forming your own personal random string of letters.   Start with the cell on the right and work left.   As they occur, write each new two-letter sequence in the space provided and shade in the corresponding subtriangle in the screen.   Continue rolling the die until all nine stage-2 addresses occur and the screen is completely filled in.   Extend the boxes in the string if needed.

Start here with the first roll.

1.	4.	7.
2.	5.	8.
3.	6.	9.

2.6B

1. Count the number of rolls needed to produce a random string that reached to each subtriangle in the stage-2 screen.

2. Compare your results to others or repeat the game a few times yourself.   Find the median for all results.   Then construct a box-and-whisker plot for the collected data.

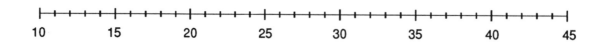

3. Use your results from 1 and 2 to predict the expected number of rolls of the die needed to reach each subtriangle in the stage-2 screen for the Sierpinski triangle.

4. Using the string of letters generated from the experiment above, list all 3-letter addresses that occurred.

5. Shade in the corresponding subtriangles that were reached using the screen provided below.

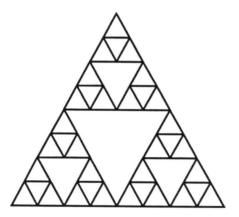

4. Find the percent of all level-3 subtriangles that were reached.

5. Scan your data to find the number of rolls necessary to reach 50% of the stage-3 subtriangles in the Sierpinski triangle.   You may have to continue rolling the die and extending the string of letters until you have enough data.

6. In reaching 50% of the stage-3 subtriangles, what percentage of the stage-4 triangles were reached?

2.7   PROBABILITY AND THE CHAOS GAME   2.7A

This activity connects the chaos game to probability through the way the letters L, T, and R are assigned to the six faces of a die.

At stage 1 the screen for the Sierpinski triangle has three subtriangles or cells with addresses L, T, and R.   What is the chance of falling in the left cell L on the first roll of the die in the chaos game?   The probability of rolling an L (1 or 2 on the die) is 1/3. The probabilities for T (3 or 4) and for R (5 or 6) are also 1/3 each.

Find the probability of ending in each cell marked on the number of rolls given.

1.    2. 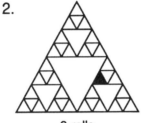   3.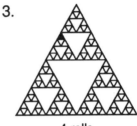

2 rolls   3 rolls   4 rolls

4.   What is the probability associated with ending in a specific stage-1 triangle in 1 roll?   a specific stage-2 triangle in 2 rolls?   a specific stage-3 triangle in 3 rolls?

Does the probability associated with each subtriangle at a given stage depend on the number of rolls of the die needed to be in it at that stage?   The answer is *no*.   For example, there are three stage-2 triangles located in each stage-1 triangle.   The sum of their probabilities is 1/9 + 1/9 + 1/9 = 1/3.   But this is exactly the probability associated with ending in the corresponding stage-1 triangle in two rolls.   Thus the probability of being in a particular stage-1 triangle after two rolls remains 1/3.

Find the probability of falling in each cell marked on the number of rolls given.

5.    6. 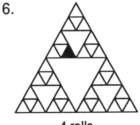   7.

3 rolls   4 rolls   5 rolls

Summary   The probability of ending in a specific stage-n subtriangle or cell for any specific number of rolls $r \geq n$ is $(1/3)^n$.

8.   As the number of rolls increases without bound, what happens to the probability of ending in a given stage-2 subtriangle at one point or another?   a given stage-n triangle for any number n?

CHANGING PROBABILITIES IN THE CHAOS GAME   2.7B

Suppose these modifications were made on assigning moves in the chaos game so that the chances of rolling an L, T, or R are not equally likely.

For 1 on the die, move halfway to L.
For 2 or 3, move halfway to T.
For 4, 5, or 6, move halfway to R.

Find the probability of ending in each cell marked using these new values.

9.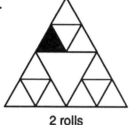
2 rolls

10.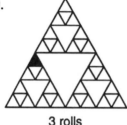
3 rolls

11.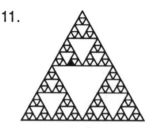
4 rolls

12. Find the probability of ending in triangle RRL after three rolls of the die. Name two other triangles that have the same probability after three rolls.

Shade in the other triangles that have the same probability of containing the point in the number of rolls given as the one already marked.

13.
2 rolls

14.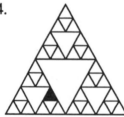
3 rolls

15.
4 rolls

16. What stage-4 triangles have probabilities of 1/48 of containing the point in four rolls?

17. What stage-5 triangle has the greatest chance of containing the point in five rolls? Find the probability.

18. What stage-6 triangle has the least chance of containing the point in six rolls? Find the probability.

19. Predict what will happen if you play the chaos game with these new conditions.

20. Simulate the results of 1500 rolls using a graphing calculator with these changes in line 25 of the code in Activity 2.2, as modified in question 2 on sheet 2.2B.
    Casio: N < .5 -> Goto 2   *TI-81:*   :If N < .5

2.8 TREES AND THE CANTOR SET

2.8A

Begin with the set of points in the closed interval [0,1]. First remove, as an open interval, the middle third of that interval. Then repeat the process over and over, each time on twice as many intervals as the last. If the iterations are repeated over and over without end, the points remaining form the *Cantor set*. This final stage consists of a set of disconnected points, distributed in cluster-like fashion over the interval.

1. The first stage in constructing the Cantor set is shown. On successive lines below it, construct the second, third, and fourth stages.

Stage 0
Stage 1
Stage 2
Stage 3
Stage 4

2. How many subintervals are in the fifth stage? in the *n*-th stage?

Let us next use trees to develop a scheme to address the separate points that remain in the final stages of the Cantor set. Suppose the tree grows such that it eventually spans the entire interval of the Cantor set. Give it specific dimensions such that it replicates the subdivision process that generates the desired Cantor set over the interval [0,1].

This tree branches off at 45° angles, with new branches 1/3 the length of the previous ones.

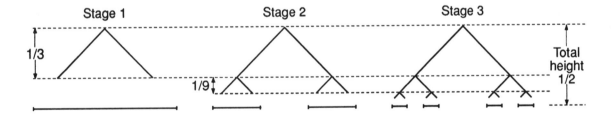

The vertical height of the tree is 1/3 at stage 1, 4/9 at stage 2, and 13/27 at stage 3.

Stage	Height
1	1/3
2	1/3 + 1/9 = 4/9
3	1/3 + 1/9 + 1/27 = 13/27
.	
.	
Final	1/3 + 1/9 + 1/27 + 1/81 + . . . = 1/2

3. Use your knowledge of series to prove that the total vertical height of the final tree, when completely grown, is 1/2.

4. Use your knowledge of geometry to prove that the total width spanned by the final tree, when completely grown, is 1.

5. Discuss the self-similarity properties of the tree at its final stage. What parts of the completed tree are strictly self-similar?

Each path in the final stage of the tree can be addressed using an endless string of the letters L and $\bar{R}$. The two endpoints of the initial interval can be named using the addresses $\overline{LLL}$ ... and $\overline{RRR}$... The bars over the letters indicate they repeat without end.

6. Use the letters L and R to give the addresses of the two endpoints for the two stage-2 intervals in the construction of the Cantor set, assuming the tree is fully grown.

7. Use the letters L and R to give the four pairs of addresses for the endpoints of the stage-3 intervals.

8. Do the endpoints of every subinterval remain as part of the final Cantor set?

Every endless string of the letters L and R correspond to a path through the completed tree. Every path through the tree corresponds to a specific point in the Cantor set. Hence, the letter strings address the points in the Cantor set.

The string $LR\overline{LR}$... names a path in the tree to a specific point in the Cantor set.

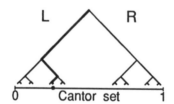

By modifying the growth rate of the tree, an interesting match to the binary decimals can be shown.

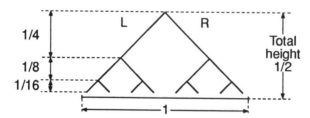

2.8C

9. Prove that this tree, when fully grown, has a height of 1/2 and a width of 1.

Associate the digit 0 with each branch L to the left. Associate the digit 1 with each branch R to the right. Now each letter-string address matches a binary decimal.

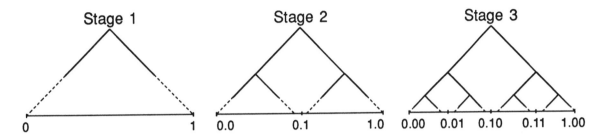

Stage 1   Stage 2   Stage 3

The stage-1 branches identify every path from $\overline{LLL}$... to $\overline{RRR}$... This corresponds to every binary decimal from $0.00\overline{0}$... = 0 to $0.11\overline{1}$... = 1.

The same closed interval [0,1] is spanned with the branches in stage 3.

	Letter-string address	Binary decimal
Stage 3	$LL\overline{L}$... to $LL\overline{R}$...	$0.00\overline{0}$... to $0.00\overline{1}$... = 0.01
	$LR\overline{L}$... to $LR\overline{R}$...	$0.01\overline{0}$... to $0.01\overline{1}$... = 0.10
	$RL\overline{L}$... to $RL\overline{R}$...	$0.10\overline{0}$... to $0.10\overline{1}$... = 0.11
	$RR\overline{L}$... to $RR\overline{R}$...	$0.11\overline{0}$... to $0.11\overline{1}$... = 1.00

Give the binary decimal address between 0 and 1 for each string of letters.

10. $RLR\overline{R}$...   11. $LLRLL\overline{L}$...   12. $LRRRLL\overline{L}$...

13. What conclusions can be drawn from this tree concerning the cardinality of the Cantor set?

We can now see that the cardinality of the Cantor set must be the same as the cardinality of the unit interval [0,1].
• Each point in the interval has a binary expansion.
• Each binary expansion corresponds to a path in the binary tree for binary decimals.
• Each such path has a corresponding path in the tree for the Cantor set.
• Each path in the tree of the Cantor set identifies a unique point in the Cantor set. Therefore, for each number in the interval, there is a corresponding point in the Cantor set. For different numbers there are different points. Thus, the cardinality of the Cantor set must be at least as large as the cardinality of the interval. On the other hand, it cannot exceed this cardinality, because the Cantor set is a subset of the interval. Therefore, both cardinalities must be the same.

2.9   TREES AND THE SIERPINSKI TRIANGLE   2.9A

In this concluding enrichment activity, trees and the Cantor set are tied to the Sierpinski triangle and the chaos game.

Imagine constructing a Cantor-like set where the intervals at each stage are separated into eight equal parts and the third and sixth parts removed.   Here the iterative procedure replaces each interval with three smaller ones.

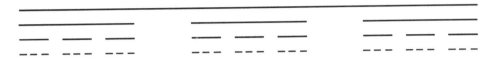

Can a tree be constructed that will generate the final set when fully grown?   The answer is *yes*.

Use three branches at each point, with the two outside branches moving off at 45° angles in opposite directions.   Make each new branch 1/4 that of the previous one from which it grows.   For the tree to have a final width of 1, it must have a total height of 1/2.   The initial height at stage 1 needs to be 3/8.

$$3/8 + 3/32 + 3/128 + 3/512 + \ldots = 1/2$$

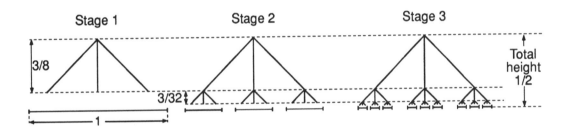

As the tree grows to its final stage, it spans the whole interval [0,1].   However, at every finite stage, only selected subintervals are spanned by the separate branches.

1. How many subintervals are spanned at stage 3?   at stage 4?   at the *n*-th stage?

2. As the tree grows without end, what happens to the number of subintervals?   What happens to their widths?

3. Discuss the self-similarity properties of the tree in its final stage.   Is the set of leaves on the fully grown tree strictly self-similar?

Different paths down the completed tree can be addressed with different endless strings of the letters L, T, and R.

4. Give the three pairs of addresses that locate the endpoints of the stage-2 subintervals spanned in the construction of this Cantor set.

5. Do the endpoints of every subinterval spanned at every stage remain as part of the final Cantor set?

It may appear that this tree, with its address strings, corresponds to the Sierpinski triangle and thus to the chaos game. Unfortunately, such is not the case.

Is there a modification that can be made to the Sierpinski triangle and the chaos game so that there is an exact correspondence to the tree we have already shown in this activity?   The answer is *yes*.   Construct a modified Sierpinski triangle by trisecting the sides, keeping only the corner subtriangles as shown.   Several stages are matched here to their corresponding trees.

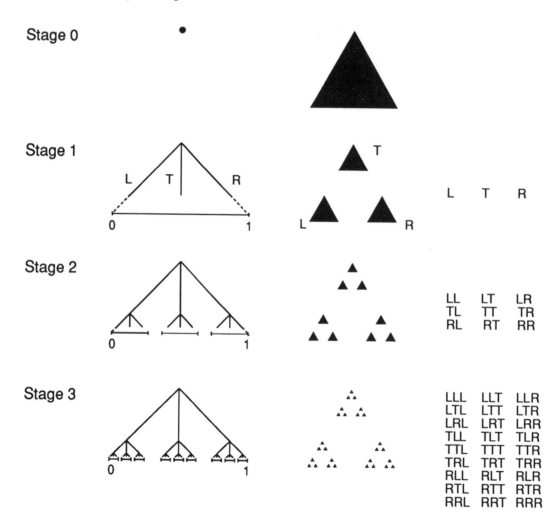

6. Here is a mapping between stage-2 points in the Cantor set and the modified Sierpinski triangle for the address RT. Do the same for LR, RL, and TT.

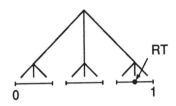

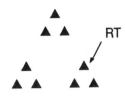

7. Given the address TRRR̄... , find the point in the Cantor set and its location on the triangle.

Tree   Address   Triangle

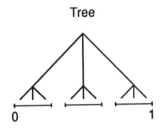

$$T\overline{R}$$
_____

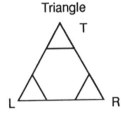

8. Given the point shown in the Cantor set below, find its address and its location on the triangle.

Tree   Address   Triangle

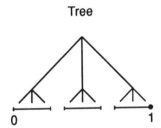

_____

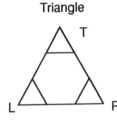

9. Given the location shown on the triangle below, find its address and point in the Cantor set.

Tree   Address   Triangle

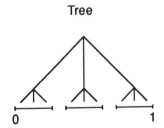

_____

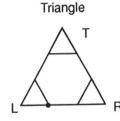

It is a simple matter to change the rules of the chaos game such that it produces this modified Sierpinski triangle. Instead of moving halfway to the vertex identified by each roll of the die, move two-thirds of the way. This sets up an iterative process that always forces the point at any stage into one of the many sets of three separated, smaller corner triangles generated at the next stage.

10. Locate a starting point in a triangle. Measure and trace the path using the two-thirds rule for the sequence LLRT. Remember, T is rolled first, and then R, followed by L and L again.

As a final question, can a tree be constructed that does corresponds directly to the original Sierpinski triangle and the original version of the chaos game? Again, the answer is *yes*. The first three stages of one possible tree are show here, drawn on top of the Sierpinski triangle. Notice that at each stage, the endpoints are located in the middle, *inside* the subtriangles. These same endpoints become the branching points for the next stage and, at that stage, lie *outside* the newly formed subtriangles.

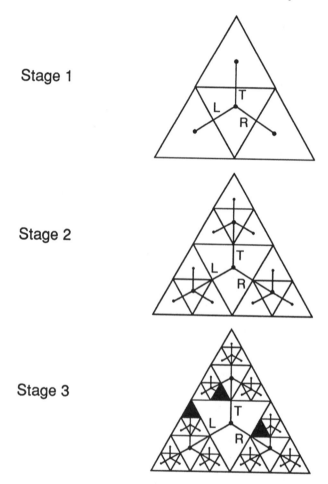

Stage 1

Stage 2

Stage 3

11. Follow the branches and give the addresses for the shaded cells in the stage-3 tree shown.

12. Let each branch in stage 1 of the tree have a length of 1. Find the total length of all branches at stage 2. at stage 3. at stage n.

13. As this tree keeps growing, will the branches ever touch each other?

Unit 3
Complexity

KEY OBJECTIVES, NOTIONS, and CONNECTIONS

Fractals are highly detailed, complex geometric shapes. One measure of their complexity is fractal dimension. By forming linear, semilog, and double log plots of data obtained by counting shaded boxes in a grid, fractal dimension is shown to follow a power law. Self-similarity dimension is also introduced as another method for computing fractal dimension when the object possesses self-similarity.

Connections to the Curriculum

In using these strategic activities as an integral part of instruction in a contemporary classroom, it is useful to know some of the points where the following activities connect to the existing curriculum. The activities on dimension have the following connections:

PRIMARY CONNECTIONS:

Power Function
Logarithms

SECONDARY CONNECTIONS:

Data Plotting
Slope
Linear Function
Exponential Function
A Limiting Concept
Geometric Patterns

Underlying Notions

Linear Function

A linear function has a straight line graph and an equation of the form $y = mx + b$.

Exponential Function

In an exponential function the independent variable appears in the exponent. Exponential functions have the form $y = ka^x$.

Power Function

Any function which has an equation of the form $y = kx^n$ is a power function.

Linear Plot

When data is plotted on linear graph paper, the grid size or scales on both axes remains constant.

Semilogarithmic Plot   A semilogarithmic graph plots the independent variable on the horizontal axis versus the log of the dependent variable on the vertical axis.

Double Logarithmic Plot   A double log plot graphs the log of the independent variable on the horizontal axis versus the log of the dependent variable on the vertical axis.

Box Dimension   One measure of the complexity of a figure is obtained by measuring the slope of a double-logarithmic plot of the number of boxes in a grid containing a portion of the figure versus the reciprocal of the scale of the grid.

Self-Similarity Dimension   Many objects which are strictly self-similar can be obtained by a recursive process of scaling and substitution in which the scale factor and the number of pieces substituted are always the same. Self-similarity dimension is the quotient of the log of the number of pieces in a replacement step and the log of the reciprocal of the scale.

MATHEMATICAL BACKGROUND

The Bigger Picture

Objects and processes which occur naturally oftentimes exhibit features which are highly complex. Fractal dimension offers one means for modeling and comparing the relative levels of such complexity. The greater the complexity of an object, the higher its fractal dimension in comparison to that of a less complex object.

To a limited extent, fractal dimension may be thought of as measuring the irregularity of an object. For example, imagine a perfectly flat piece of tinfoil of infinite length and width. In its planar state this foil would be described as having dimension 2. However, upon progressively gathering the tinfoil and crushing it, the complexity of the surface would increase, and a three-dimensional ball would be formed. In short, as the foil deforms from its flat state to a solid ball, the surface grows in complexity as it becomes increasingly more twisted, bent, and rugged. As an idealized measure of this complexity, the fractal dimension would increase from 2 to 3.

Fractal shapes can generally be separated into two basic categories, deterministic and nondeterministic.

• Often deterministic fractals are formed by repeating a principle geometric pattern over all scales. Strictly self-similar objects are in this class. Examples of such fractals include the Koch curve and the Peano curve.

• Nondeterministic fractals involve some random variation in the construction process, but nonetheless, the way that details appear under continuous magnification seems to be invariant. This type of fractal can be used to model coastlines and mountains.

We can measure the way in which details relate to the scale of magnification in several ways:

- compare the number of replacement parts at each level of a deterministic construction process to the relative size of the replacement parts, a measure that leads to self-similarity dimension.

- compute a ratio involving the relative size of the boxes and the number of boxes of a given size required to cover the fractal shape. The measure offered by this ratio is referred to as box dimension.

- compare the number of rulers required to circumscribe perimeter to the relative length of the ruler, a measure particularly suitable for studying coastlines. Although this method is not practiced in this volume, it is similar to the box-counting method in that instead of counting blocks one counts line segments of different scales.

- employ spectral analysis by computing the slope of a power spectrum. Like the preceding method, this approach to measuring complexity is not practiced in this volume. However, it is herein mentioned to indicate that alternatives exist to the other methods for use in certain advanced applications.

Despite the fact that self-similarity dimension can generally be obtained from a figure very rapidly, this measure only applies to deterministic figures possessing strict self-similarity. However, box dimension can be computed for figures in one, two, three, or even higher dimensional spaces irrespective of whether or not the objects exhibit self-similarity. In order to compute box dimension in two dimensions, count the squares that cover the object from grids of differing scales. As the grid size decreases, the number of boxes required to cover the fractal shape grows according to a power law. The exponent in this power law is the box dimension, a real number which characterizes the complexity of the fractal shape. For a point, line, and disk, one obtains the expected Euclidean dimensions of 0, 1, and 2, respectively by the boxcounting approach.

Fractal dimensions are computable by a number of approaches as described above and seek to characterize the complexity of an object. In extreme cases one may obtain different results by different approaches to fractal dimension. Moreover, fractal dimension may not well represent structures that are composites of two or more parts having distinct dimensions. In such cases, applying box dimension yields the dimension of the most complex part even if that part is only a small section of a larger shape having significantly less complexity everywhere else.

Growth Laws

Linear Relationship

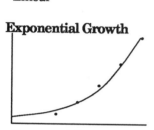

Linear

A relationship between two variables is linear if the ratio $\Delta y/\Delta x$ (slope) between a change Δy in the dependent variable to a change Δx in the independent variable remains constant throughout the domain of the independent variable. In such a case, the relationship may be modeled by an equation of the form $y = mx+b$. Graphing data points (x, y) from such a relationship on standard graph paper where both the horizontal and vertical scales are constant leads to a linear graph.

Exponential Growth

Exponential

An exponential relationship exists between two variables when the relationship may be modeled by an equation of the form $y = ka^x$. When data points (x, y) from such a relationship are graphed on standard graph paper with constant horizontal and vertical scales, a curve results and not a straight line (exceptions: $a=0$ and $a=1$). However, the following derivation shows that if an exponential relationship exists between two variables x and y, then by plotting points of the form $(x, \log y)$ on standard linear graph paper, a linear graph will appear.

Let $a > 0$ and

$$y = ka^x$$

Then

$$\log y = \log ka^x$$
$$= \log a^x + \log k$$
$$= x \log a + \log k$$

Notice that this last equation is of the linear form $Y = AX+K$. Slope A equals the log of the base a in the original exponential function, while the y-intercept K equals the log of the original coefficient k.

Power Law

Power

A power relationship exists between two variables when the relationship may be modeled by an equation of the form $y = kx^r$. When data points (x, y) from such a relationship are graphed on standard graph paper with constant horizontal and vertical scales, a curve results and not a straight line. The one exception is when r is equal to 0 or 1. However, the following derivation shows that if a power relationship exists between two positive variables x and y, then by plotting points of the form $(\log x, \log y)$ on standard graph paper, a linear graph will appear.

Let

$$y = kx^r$$

Then

$$\log y = \log kx^r$$
$$= \log x^r + \log k$$
$$= r \log x + \log k$$

Notice that this last equation is of the linear form $Y = rX+K$. The slope r equals the exponent in the original power function, while the y-intercept K equals the log of the coefficient k in the power formula.

Graphing

We remark that the above derivations work regardless of the base of the logarithms used. However, unless noted otherwise we assume that in the Activity Sheets logarithms with respect to base 10 are used.

By plotting data on standard (constant scale) graph paper we can detect if a linear, exponential, or power relationship exists.

- If plotting the original data points (x, y) gives a linear graph, then by reading the slope m and y-intercept b from the graph indicates that the underlying linear relationship has the equation $y = mx + b$.

If this plot of the data points does not reveal a linear relationship, then the following tests may be performed to detect an exponential or power law.

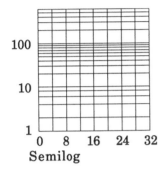

Semilog

- If plotting transformed data points $(x, \log y)$ gives a linear graph, then the underlying relationship is exponential with base equal to the antilog of the slope and coefficient equal to the antilog of the y-intercept.

- If plotting transformed data points of the form $(\log x, \log y)$ yields a linear graph, then an underlying power relationship exists.

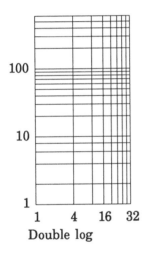

Double log

For convenience, graph paper has been published which obviates the need to actually transform the variables into logarithmic form for the semilog or double log plots. Semilog graph paper is used when one suspects that an exponential law relates the variables. On such paper the vertical axis is already marked in a logarithmic scale. Double log paper is used when testing data for a power relationship. On double log paper, both the horizontal and the vertical axes have been marked according to logarithmic scales. Plotting points of the form (x, y) on paper with these transformed scales already present is equivalent to but more convenient than transforming the data using logarithms and then plotting the results on standard linearly scaled paper. Even more conveniently, some graphing calculators even allow the entry of data points (x, y) and then simplified automatic graphing of either linear, semilogarithmic, or double logarithmic plots.

Fractal Dimensions

Box Dimension

The data gathered in this packet for plotting will be obtained from grids placed over a figure to be measured for its level of complexity. For various grid sizes, the number of boxes y in each grid which contain a portion of the figure will be plotted against the reciprocal of the scale x which determined the size of the grid. By graphing the resulting points using linear, semilogarithmic, and double logarithmic plots, we will discover that the relationship between box count y and grid size $1/x$ only gives a linear graph when $\log y$ is plotted against log $1/x$. Accordingly, the relationship between box count and scale is a power relationship. The slope of the line obtained in this double log plot is the exponent in the power relationship. It specifically represents the fractal dimension of the figure, and it will be 1 for a smooth line, 2 for a solid disk or square, and between 1 and 2 for fractal curves.

Self-Similarity Dimension

One alternative approach to box dimension for measuring the complexity of a figure can be applied in certain cases in which the figure is strictly self-similar. Self-similar objects, such as the Sierpinski triangle or the Koch curve, are obtained by a recursive process of scaling and substitution in which the scale factor and the number of pieces substituted are always the same. Self-similarity dimension is the quotient of the log of the number of pieces in a replacement step and the log of the inverse of the scale. It is important to stress that the self-similarity dimension characterizes the limit-object and not any of the intermediate steps in the construction process. In fact, this notion can even be implemented if parts of the figure overlap each other, but when this is the case the self-similarity dimension will not necessarily equal the box dimension.

Additional Readings

Fractals for the Classroom, Chapter 4.

USING THE ACTIVITY SHEETS

3.1 Construction and Complexity

Specific Directions: In addition to a ruler, three pens containing different colored ink will be helpful, though not necessary.

An equilateral triangle is exhibited on dot paper. The construction requires that *each* side be replaced by 4 segments as shown on the activity sheet. The resulting figure consists of 12 segments which should in turn be subjected to the same replacement process. Using a distinct color to identify each stage of the construction process simplifies completion of the next stage. The construction process should be repeated three times.

Implicit Discoveries: The complexity of the figure increases with each application of the construction process. An implicit question arises as to how the complexity might be measured if the process was allowed to continue ad infinitum to the final state of the figure. This question is addressed in the remainder of this unit.

3.2 Fractal Curves

Specific Directions: In addition to a ruler, three pens containing different colored ink will be helpful, though not necessary.

The construction process defined on Sheet 3.2A is made easier by changing the color of the ink after each stage of the drawings.

Extensions: Design your own replacement patterns and draw three stages of the resulting curves.

3.3 Curve Fitting

Specific Directions: You will need a ruler and a copy of each of the sheets of graph paper supplied on Sheets 3.3C, 3.3D, and 3.3E.

As detailed in Exercises 1 and 2 on Sheets 3.3A and 3.3B, construct three graphs (linear, semilogarithmic, and double logarithmic) for each of the three supplied data sets thereby finishing with a total of nine distinct graphs. Then complete exercises 3-6 on Sheet 3.3B in order to discover how underlying relationships existing between two variables can be detected within collected data.

Implicit Discoveries: By utilizing graph paper designed specifically for the purpose of analysing underlying relationships between two variables, it is possible to detect if the variables are related linearly (speedometer data), exponentially (population data), or according to a power function (skydiver data). In the case of a power relationship, plotting the data on double logarithmic paper yields a line having a slope equal to the exponent in the power function.

Extensions: Use linear, semilogarithmic, and double logarithmic plots to analyze data from a current almanac.

3.4 Curve Fitting Using Logs

Specific Directions: You will need a ruler, two sheets of standard graph paper, and a scientific calculator. Do not use copies of the sheet from Activity 3.3C, because the scaling is different here.

Plot the five graphs required in exercises 1-5. Then follow the steps prescribed in exercise 6 to demonstrate that, if a line closely fits a plot of the points $(x, \log y)$, then the underlying relationship existing between variables x and y must be exponential. Similarly, follow the steps prescribed in exercise 7 to show that, if a line closely fits a plot of the points $(\log x, \log y)$, then an underlying power relationship must exist between variables x and y.

Implicit Discoveries: By using logarithms to transform data, it is possible to detect if two variables are related linearly (speedometer data), exponentially (population data), or according to a power function (skydiver data). In the case of a power relationship, plotting the data in the form $(\log x, \log y)$ yields a straight line having a slope equal to the exponent in the power function.

Extensions: Analyze data from scientific experiments by utilizing linear plots (x versus y), semilogarithmic plots (x versus $\log y$), and double logarithmic plots ($\log x$ versus $\log y$).

3.5 Curve Fitting Using Technology

Specific Directions: This Activity requires the use of a graphing calculator or a computer package which permits plotting data and fitting selected curves to the resulting plots. For illustrative purposes, detailed examples are given for the TI-81 graphing calculator.

The Pearson Correlation Coefficient r can be used to measure the goodness of fit of a particular curve to the data plot. The closer the value of r is to 1, the better the fit.

Extensions: Test bivariate data sets from laboratory experiments for underlying linear, exponential, or power relationships.

3.6 Box Dimension

Specific Directions: You will need a ruler and copies of the three sheets of graph paper supplied with Activity 3.3.

By viewing a given figure under a grid of boxes, it is possible to count the number of boxes which contain a portion of the shape. The initial task is to graphically see, for each such shape, that data relating the resulting boxcounts to the grid size exhibits an underlying power relationship. Box dimension may then be computed as the slope of the best fitting line in the double logarithmic plot of this data. Due to graphical imperfections and ambiguities the results of manual boxcounting can vary rather widely. Also, for more accurate results it would be necessary to use finer grid sizes. For example, using smaler grid sizes for the BLACK HOLE example the result would approach the theoretical value 2 of its box dimension.

Implicit Discoveries: The computed value of box dimension increases with greater complexity in the fractal shape.

3.7 Box Dimensions and Coastlines

Specific Directions: Use the boxcounting techniques practiced in Activity 3.6. For each map and given grid, be sure to count every box in the grid that contains any portion of the coastline within the interior of the box.

Extensions: Measure the box dimension of the coastline of Great Britain as represented by maps from more than one atlas. Why might these measured box dimensions differ?

3.8 Box Dimension for Self-Similar Objects

Specific Directions: Review the construction process used to make the Koch snowflake (Sheet 3.1A) and the 3/2 curve (Sheet 3.2A, Construction B). Then, following the instructions on Sheet 3.8A, use the boxcounting techniques practiced in Activities 3.6 and 3.7 in order to measure the box dimension of each shape supplied on Sheets 3.8B, 3.8C, and 3.8D. For each figure and given grid, be sure to count every box in the grid that contains any portion of the figure within the interior of the box.

Implicit Discoveries: The box dimension of a figure is determined by the region of the figure with the greatest complexity.

3.9 Similarity Dimension

Specific Directions: You will need a ruler, and completed versions of Sheets 3.1A and 3.2A.

Complete the two constructions provided on Sheets 3.9C and 3.9D. The three constructions presented on Sheets 3.1A and 3.2A should also be completed if they have not already been finished. At each level of a given construction be careful to replace every line segment by the indicated pattern of smaller segments. Then complete Sheet 3.9A in order to compute the similarity dimension for each of the five constructions.

Implicit Discoveries: The computed value of similarity dimension increases with increased complexity in shape of the fractal.

Extensions: Design other replacement patterns and compute the associated similarity dimension for the figures which would result if the constructions were repeated ad infinitum.

3.1 CONSTRUCTION AND COMPLEXITY

3.1A

There are many ways to form fractals that are self-similar.   These figures have the property that each part, however small, contains an exact replica of the whole.   We can generate examples by a recursive process of scaling and substitution, where both the scaling factor and the number of pieces substituted are always the same.

The following construction uses a recursive, scaling and substitution process to produce a familiar fractal image called the Koch snowflake.

Construction   Using trisection points, replace each
line segment with the pattern shown.

Repeating the iterative process starting from the line segment *ad infinitum* generates a fractal called the Koch curve. Apply the construction process three successive times for all three sides of the equilateral triangle given.   Count dots to locate the trisection points at each stage.   Remember, each segment is transformed into four shorter segments at each step in the iterative process.   Each of these shorter segments is one third the length of the segment being replaced.   Repeating the iterative process *ad infinitum* generates the Koch snowflake. Observe that the final object is made of three pieces each of which is a complete Koch curve.

1. Count the number of segments at stages 0, 1, 2, and 3. Discover and extend the number pattern through the first five stages of growth. Generalize to find the number of segments for level n.

Stage	0	1	2	3	4	5 ...	n
Segments	3						

2. Imagine repeating the process without end. Visualize and describe how the figure changes.

3. In moving from one stage to the next, the snowflake grows in perimeter. Let each side of the initial equilateral triangle be 1 and find the perimeter at each of the first five stages of growth. Generalize to find the perimeter for level n.

Stage	0	1	2	3	4	5 ...	n
Perimeter	3						

4. Let the iteration process continue *ad infinitum*. What is the perimeter of the resulting Koch snowflake?

5. As the snowflake grows, its area also increases. Starting from an initial side of 1, find the area at each of the first five stages. Leave answers in radical form. At stage 0 the area is $\sqrt{3}/4$.

6. Generalize to find the area for level n. What is the area of the completed Koch snowflake?

7. Is the perimeter of the Koch snowflake finite or infinite? Is its area finite or infinite?

As the construction of the snowflake process proceeds from stage to stage, the areas of the resulting figures converge to a finite value. At the same time, the perimeter diverges, growing large without bound. The shapes generated from this iterative construction increase in complexity as they continue from stage to stage toward the final state, the fractal known as the Koch snowflake. It is composed of three pieces each of which is a Koch curve.

8. Is the Koch curve self-similar? Is it strictly self-similar? Is the Koch snowflake self-similar?

Koch curve construction

3.2   FRACTAL CURVE   3.2A

Each of the two construction processes below generates a fractal through scaling and substitution.

Construction A   Replace each line segment with the pattern shown. Carry the construction through stage 3.

Construction B   Replace each line segment with the pattern shown. Carry the construction through stage 2.

1. Apply construction A through stage 3 and construction B through stage 2.

2. If the length of the initial segment is 1 in each case, what is the segment length at stage 3 for construction A? at stage 2 for construction B?

The scaling factor is the ratio of the lengths of new segments to those they replace.

3. What is the scaling factor for construction A?
   What is the scaling factor for construction B?

4. What is the scaling factor used in the construction of the Koch snowflake / Koch curve?

Compare the Koch curve to those just constructed. All constructions started with one line segment at stage 0.   As the figures moved through their stages of development, they became more and more complex.

5. Which of the three curves appears to be more complex at stage 1?   at stage 2?

6. These iterative construction algorithms all produce fractals when repeated *ad infinitum* .   All three fractals have strict self-similarity.   At their limiting states, will the Koch curve, right-angled curve A, or right-angled curve B appear most complex?

Both the Koch snowflake / Koch curve and the right-angled curves are highly complex and contain a great deal of detail, even at relatively early stages of their growth.   As they move toward their limiting states, their complexity increases.   The remaining activities of this chapter are devoted to establishing means for measuring this complexity.

3.3   CURVE FITTING   3.3A

One measure of the complexity present in a geometric representation of a fractal shape involves a graph of data obtained from the shape itself.   This activity sheet focuses on the necessary graphing skills.

DIRECTIONS   Use the data supplied in Tables A, B, and C for the experiments that follow. You will need a ruler and three types of graph paper.

> Standard graph paper
> Semilogarithmic graph paper
> Double logarithmic graph paper

A. In order to check the accuracy of his speedometer, the driver of a vehicle traveling westward on Interstate Highway 70 across Kansas sets his cruise control at exactly 65 mph. While passing mileposts, the driver records the following elapsed minutes and accumulated distances in tenths of a mile.

TABLE A	Minutes x	6	10	15	25	31
	Distance y	65	108	163	271	336

B. The 1990 World Almanac supplies data on the world population since 1650.   Here are the figures, in tens of millions, over five-decade intervals from 1650.

TABLE B	Decades x	5	10	15	25	30
	Population y	63	71	91	160	250

C. A skydiver is plummeting head first in a state of free-fall.   The distance she falls is measured vertically from the initial jumping point in intervals of tenths of a second.

TABLE C	Seconds x	10	15	20	25	30
	Feet y	16.0	36.2	64.5	100.6	144.9

1. Use the data in Table A to plot the points (x, y) on three different graph papers.

> Standard graph paper
> Semilogarithmic graph paper
> Double logarithmic graph paper

In each case, describe the basic trend as you see it by drawing the best fitting straight line or smooth curve through the points.   Label each line or curve "speedometer."

2. Repeat the process for the data in Tables B and C. Plot your points on the same sheets of graph paper used above. Label the second set of graphs "population" and the third set "skydiver."

When the points (x, y) form a straight line on standard graph paper, a linear relationship is indicated between the variables. The slope of the straight line is the coefficient of the variable x.

When the points (x, y) form a straight line on semilogarithmic paper, an exponential relationship is indicated between the variables. The slope of the straight line equals the logarithm of the base in the exponential function.

When the points (x, y) form a straight line on double logarithmic paper, a power relationship is indicated between the variables. The slope of the straight line equals the exponent in the power function.

FUNCTION	FORMULA	GRAPH PAPER	SLOPE
Linear	$y = mx + b$	Standard	m
Exponential	$y = ka^x$	Semilog	$\log a$
Power	$y = kx^m$	Double log	n

3. Compare the three "speedometer" graphs. On which graph paper did the data of Table A appear as a straight line? What does this tell you about the relationship between the time and distance variables for a vehicle moving at a constant velocity?

4. Compare the three "population" graphs. On which graph paper did the data Table B appear as a straight line? What does this tell you about the relationship between time and world population?

5. Compare the three "skydiver" graphs. On which graph paper did the data of Table C appear as a straight line? What does this tell you about the relationship between time and distance variables for the skydiver's descent?

6. Compute the slope by using your ruler to measure the rise and run of the graph that gave a straight line for the skydiver data.

STANDARD GRAPH PAPER

3.3C

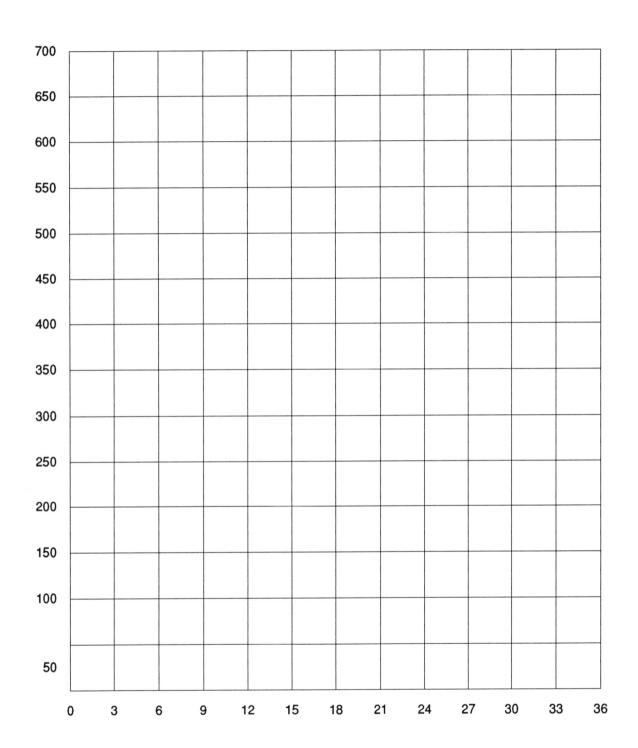

SEMILOGARITHMIC GRAPH PAPER

3.3 D

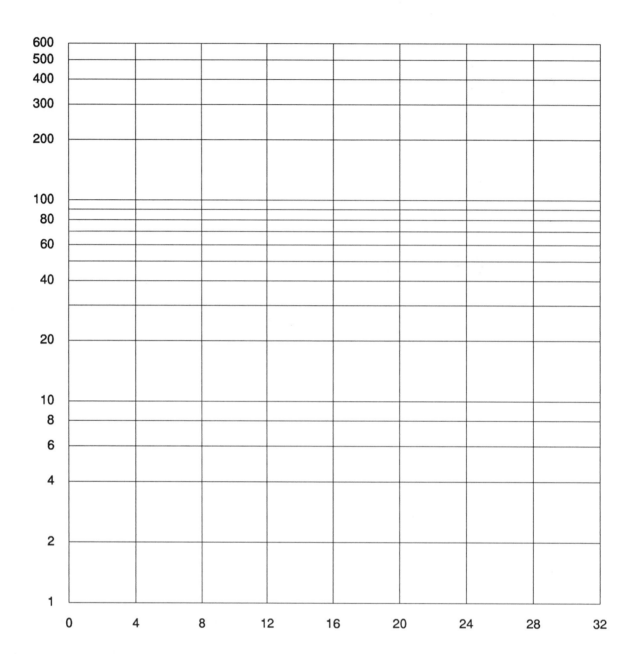

DOUBLE LOGARITHMIC GRAPH PAPER

3.3E

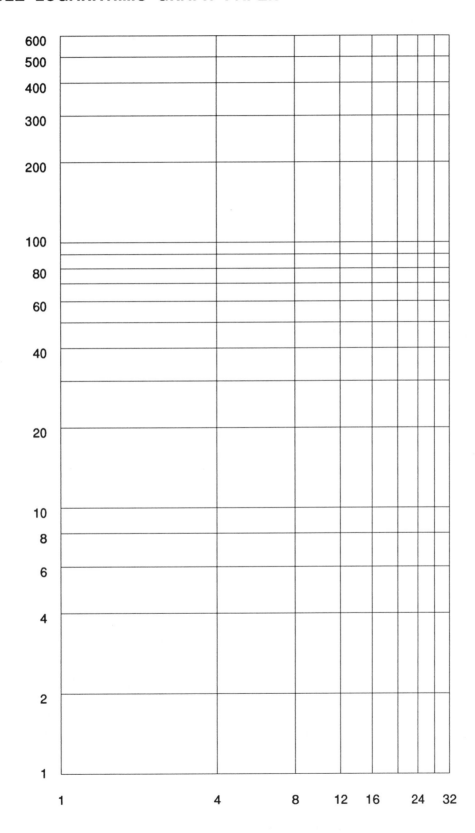

3.4 CURVE FITTING USING LOGS

One approach to measuring the complexity present in a geometric representation of a fractal shape depends upon the relationship between a function and its graph.

DIRECTIONS   Use standard graph paper, a ruler, and a scientific calculator. Each activity below relates to functions in three forms.

LINEAR	EXPONENTIAL	POWER
$y = mx + b$	$y = ka^x$	$y = kx^r$

On a single piece of graph paper, draw three separate axes using the domain [1,6].

1. Sketch graphs of the following three functions.

    a. Linear function $y = 3x + 1$
    b. Exponential function $y = 0.25 \cdot 2^x$
    c. Power function $y = 0.1 \cdot x^3$

Follow the directions in exercises 2-5 to plot two additional graphs. Use logarithms with base 10.

2. Complete the table for the exponential function $y = 0.25 \cdot 2^x$.

x	1	2	3	4	5	6
y						
log y						

3. Plot x versus log y obtained from the exponential function $y = 0.25 \cdot 2^x$.

4. Complete the table for the power function $y = 0.1 \cdot x^3$.

x	1	2	3	4	5	6
y						
log x						
log y						

5. Plot log x versus log y obtained from the power function $y = 0.1 \cdot x^3$.

6. Use these steps to prove that, by plotting x versus log y for the exponential function $y = ka^x$, a linear graph will result having the form $Y = AX + K$.

  a. Take the logarithm of the expressions on both sides of the equation and simplify to the linear form log $y = x$ log a + log k.

  b. Substitute Y for log y, A for log a, K for log k, and let $X = x$. Write the resulting equation.

  c. How can the constants in the original exponential equation be determined from the constants in the associated linear form?

7. Use these steps to prove that, by plotting log x versus log y for the power function $y = kx^r$, a linear graph will result having the form $Y = rX + K$.

  a. Take the logarithm of the expressions on both sides of the equation and simplify to the linear form log $y = r$ log x + log k.

  b. Substitute Y for log y, X for log x, and K for log k. Write the resulting equation.

  c. How can the constants in the original power equation be determined from the constants in the associated linear form?

8. In each of the above exercises, common logarithms, base 10 were indicated.

  a. The linear graph formed in exercise 3 was obtained from an exponential function where x was plotted against log y. Would the graph have been nonlinear if natural logarithms, base e had been used instead? Explain your conclusion.

  b. The linear graph formed in exercise 5 was obtained from a power function where log x was plotted against log y. Would the graph have been nonlinear if natural logarithms, base e had been used instead? Explain your conclusion.

3.5 CURVE FITTING USING TECHNOLOGY   3.5A

This activity provides a quick way of assessing whether or not a linear, power, or exponential function, among others, is a good fit to a data set. Follow these steps.

- Set appropriate range parameters
- Enter the statistics mode
- Enter the data set
- Create a scatter plot
- Test the fit of the selected function to the data set
- Choose the best fit based on the correlation value *r*

Seconds	Feet
X	Y
1.0	16
1.5	36.2
2.0	64.5
2.5	100.6
3.0	144.9

Take another look at the data for the skydiver in *Activity 3.3A*. Recall that this involves a free-fall after leaving the airplane. The data recorded in the table at the right represents the distance the person had fallen, *y* (in feet) after an elapsed time, *x* (in seconds).

The general procedure for treating the data can be executed on a variety of graphing calculators and using selected software packages on microcomputers. The following 20 key strokes provide the essential steps for using the *TI-81* graphing calculator. At *Step 10*, make the choice among a linear, exponential, and power function.

Visually scan the data set and set range parameters ....
    1.   2nd MATRX
Select DATA menu
    2.   2:ClrStat
    3.   ENTER
    4.   2nd MATRX
Return to DATA menu
    5.   1:Edit
Enter data set
    6.   2nd MATRX
Select Draw menu
    7.   2:Scatter
    8.   ENTER
Return to STAT CALC menu to test linear function
    9.   2nd MATRX
    10.   2:LinReg, 4:ExpReg, or 5:PwrReg
    11.   ENTER
    12.   Y =
    13.   CLEAR
    14.   VARS
Select LR menu
    15.   4:RegEQ
    16.   ENTER
    17.   GRAPH
    18.   2nd MATRX
Select Draw menu
    19.   2:Scatter
    20.   ENTER

```
RANGE
Xmin=0
Xmax=3
Xscl=.1
Ymin=0
Ymax=150
Yscl=10
Xres=1
```

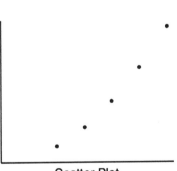

Scatter Plot

3.5B

To test for the fit to a linear, exponential, or power function, make the proper selection in *Step 10*.

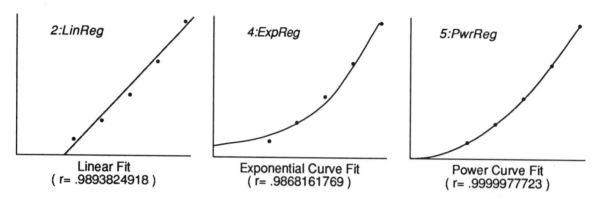

| Linear Fit | Exponential Curve Fit | Power Curve Fit |
| (r= .9893824918) | (r= .9868161769) | (r= .9999977723) |

The power function appears to fit the data best and confirms what we already know about the physics of a free-falling object. Specifically, $d = 16t^2$ is the governing relationship, where d is the distance fallen, in feet, and t is the time, in seconds.

1. Enter the "skydiver" data and compare your results with that shown above.

2. In order to check the accuracy of his speedometer, the driver of a vehicle traveling westward on Interstate Highway 70 across Kansas sets his cruise control at exactly 65 mph. While passing mileposts, the driver records the following elapsed minutes and accumulated distances in tenths of a mile.

Minutes x	6	10	15	25	31
Distance y	65	108	163	271	336

Determine which of the three functions, linear, exponential, or power, best fits the speedometer data.

3. The 1990 World Almanac supplies data on the world population since 1650. Here are the figures, in tens of millions, over five-decade intervals from 1650.

Decades x	5	10	15	25	30
Population y	63	71	91	160	250

Determine which of the three functions, linear, exponential, or power, best fits the population data.

3.6 BOX DIMENSION

3.6A

Box dimension supplies one approach to measuring the level of complexity present in a given shape. We compute the box dimension of a figure by measuring the slope of a double logarithmic plot of boxcount versus the reciprocal of the grid size.

DIRECTIONS   Use a ruler and copies of the three sheets of graph paper supplied with Activity 3.3.

Standard graph paper
Semilogarithmic graph paper
Double logarithmic graph paper

Three figures appear on sheets 6C, 6D, and 6E titled WAVE, BLACK HOLE, and FUNCTION.   Complete the steps below for each sheet.

1. Count the number of boxes in the grid that contain parts of the figure and record your result in the accompanying table.   Be sure to count every box in the grid that contains any portion of the figure within the interior of the box.

2. Make a graph of the points $(1/x, y)$ on standard graph paper.   Describe the basic trend by drawing what you see as the best fitting straight line or smooth curve through these points.

3. Make a graph of the points $(1/x, y)$ on semilogarithmic graph paper. Describe the basic trend by drawing what you see as the best fitting straight line or smooth curve through these points.

4. Make a graph of the points $(1/x, y)$ on double logarithmic graph paper. Describe the basic trend by drawing what you see as the best fitting straight line or smooth curve through these points.

Compare your three graphs.

5. On which graph are the data points best approximated by a straight line? Does a linear, exponential, or power relationship exist between scale and boxcount?

6. Use your ruler to measure the vertical rise and horizontal run of each best fitting straight line to the data points plotted.

	WAVE	BLACK HOLE	FUNCTION
Rise			
Run			

The slope of the straight line graph of the boxcount versus 1/(scale x), in a double logarithmic plot is the box dimension of the figure. Box dimension yields one measure of the complexity of a figure.

$$\text{Box dimension} = \text{Slope} = \frac{\text{Rise}}{\text{Run}}$$

7. Compute the box dimension of each figure as the slope of the line.

	WAVE	BLACK HOLE	FUNCTION
Box dimension			

8. Compare the results obtained for each figure. By studying the three figures and their associated box dimensions, what meaning might be attached to a large computed value for box dimension as compared to a smaller value for box dimension?

WAVE

scale
1/4

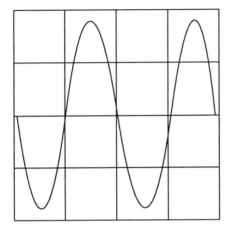

scale
1/8

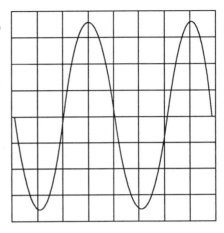

scale
1/12

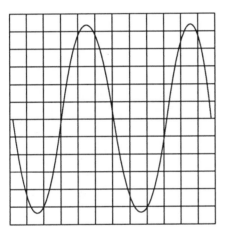

scale
1/16

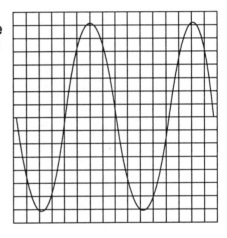

scale
1/24

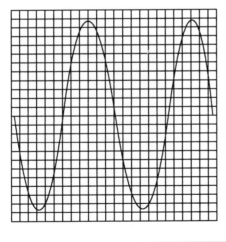

scale
1/32

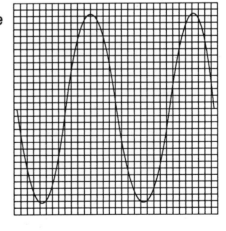

1 / (scale x)	4	8	12	16	24	32
boxcount y						135

BLACK HOLE

3.6D

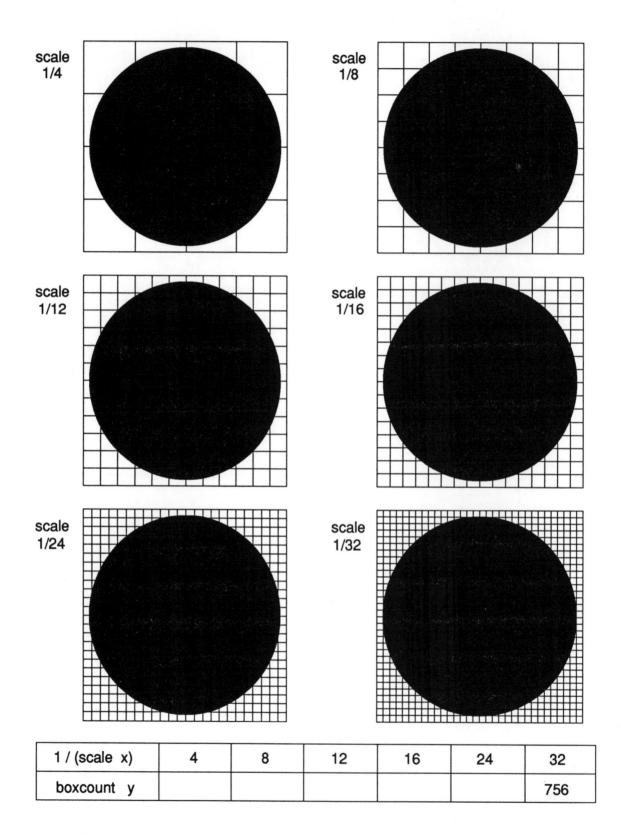

1 / (scale x)	4	8	12	16	24	32
boxcount y						756

FUNCTION

3.6E

scale
1/4

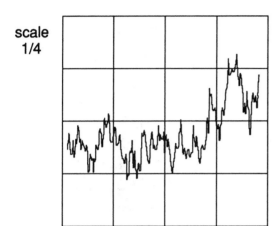

scale
1/8

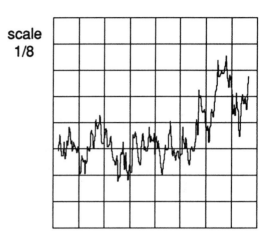

scale
1/12

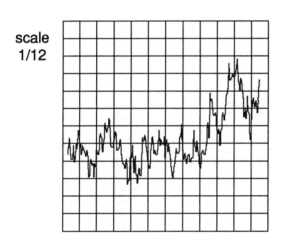

scale
1/16

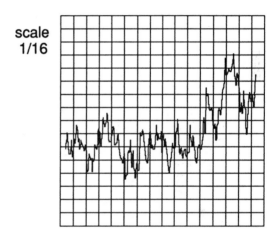

scale
1/24

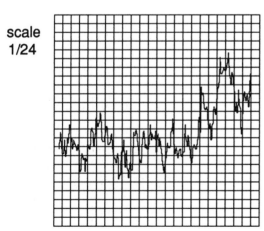

scale
1/32

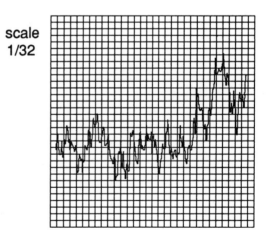

1 / (scale x)	4	8	12	16	24	32
boxcount y						150

3.7 BOX DIMENSION AND COASTLINES   3.7A

We can utilize box dimension to measure the complexity of various shapes including the complexity of coastlines.

1. For each of the six maps of Great Britain, use the boxcounting techniques practiced in Activity 3.6.   Enter the results in the table. Be sure to count every box in the grid that contains any portion of the coastline within the interior of the box.

1/(scale x)	4	8	12	16	24	32
boxcount y						283

2. Form a double logarithmic plot of the boxcounts y versus 1 / scale x. Choose one of the following approaches.

   a. Use double logarithmic graph paper as in Activity 3.3.
   b. Plot points of the form (log $1/x$, log y) on standard graph paper as in Activity 3.4.
   c. Use a computer package or graphing calculator as in Activity 3.5.

3. Compute the box dimension of the coastline by determining the slope of the best fitting line.

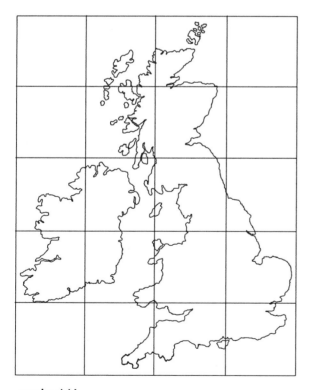

scale 1/4

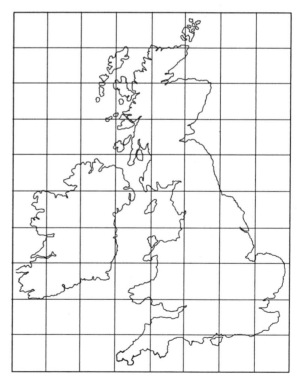

scale 1/8

3.7B

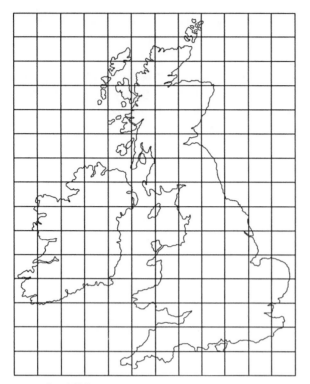

scale 1/12

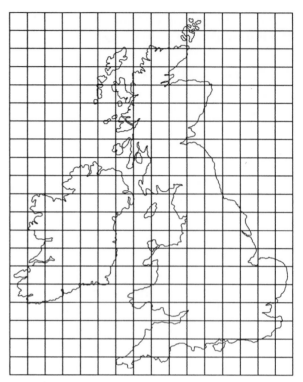

scale 1/16

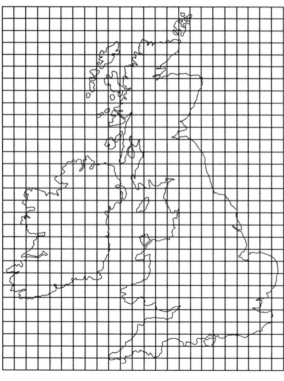

scale 1/24

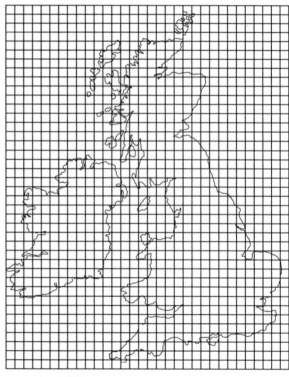

scale 1/32

3.8   BOX DIMENSION FOR SELF-SIMILAR OBJECTS   3.8A

Simlarity dimension offers a simple means for measuring complexity in a fractal shape when the figure is strictly self-similar. Details of the procedure appear in Activity 3.9. Though more difficult to apply, box dimension supplies an approach to measuring complexity irrespective of whether or not the shape exhibits self-similarity.

DIRECTIONS   Review the techniques practiced in Activity 3.7 before attempting the following experiments.

Three figures titled KOCH CURVE, 3/2 CURVE, and COMPOSITION appear on sheets 3.8B, C, and D along with their boxcount tables. Complete exercises 1-3 below first using the Koch curve.   Then repeat the exercises for each of the other two figures.

1. Count the number of boxes in the grid that contain parts of the figure and record your result in the accompanying table. Be sure to count every box in the grid that contains any portion of the figure within the interior of the box.

2. Using the data from your table as completed in exercise 1 above, form a double logarithmic plot of the boxcounts y versus the 1/(scale x). Choose one of the following approaches.

     a. Use double logarithmic graph paper as in Activity 3.3.
     b. Plot points of the form (log 1/x , log y) on standard graph paper as in Activity 3.4.
     c. Use a computer or graphing calculator as in Activity 3.5.

> The slope of the straight line graph of boxcount y versus 1/(scale x) in a double logarithmic plot is the box dimension of the figure.   Box dimension yields one measure of the complexity of a figure.

3. Compute the box dimension of the figure by determining the slope of the best fitting line.

	KOCH CURVE	3/2 CURVE	COMPOSITION
Box dimension			

4. When a figure such as the composition curve is formed from distinct parts having different box dimensions, which part will determine the box dimension of the whole figure?

5. What does your conclusion to exercise 4 imply for the measured box dimension of the coastline of Great Britain obtained on sheet 3.7A?

KOCH CURVE

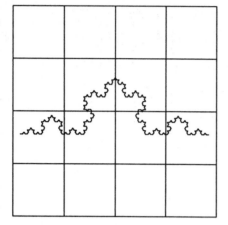
scale 1/4

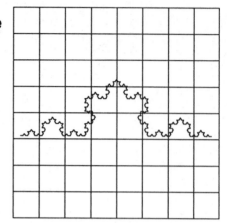
scale 1/8

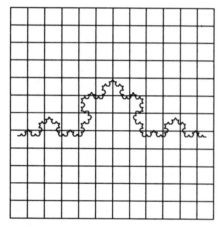
scale 1/12

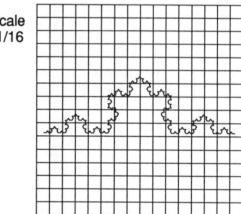
scale 1/16

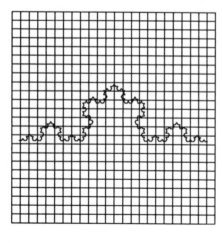
scale 1/24

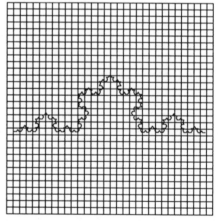
scale 1/32

1 / (scale x)	4	8	12	16	24	32
boxcount y						74

3/2 CURVE

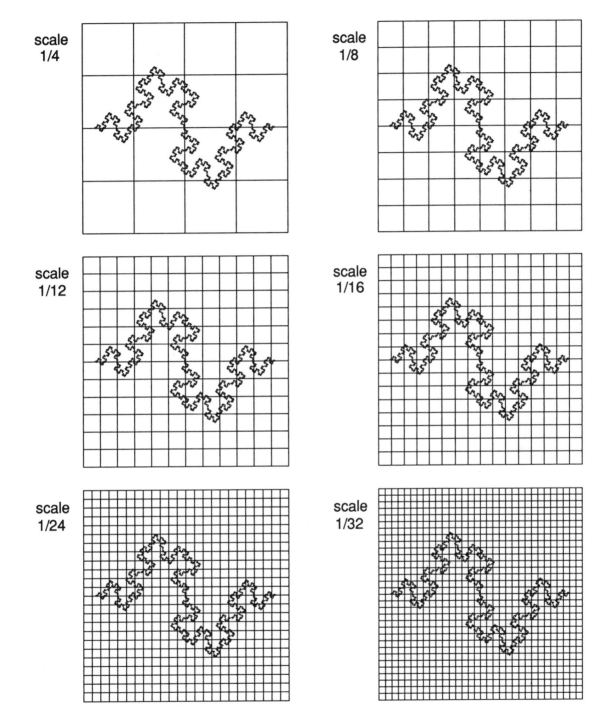

1 / (scale x)	4	8	12	16	24	32
boxcount y						155

COMPOSITION CURVE

3.8D

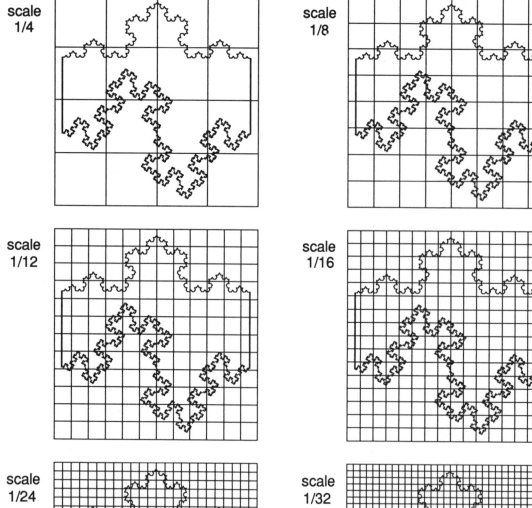

1 / (scale x)	4	8	12	16	24	32
boxcount y						255

3.9 SIMILARITY DIMENSION   3.9A

Similarity dimension applies to strictly self-similar figures.   It offers a means of measuring complexity in a fractal shape when each part of the figure is a replica of the whole.

DIRECTIONS   The five geometric constructions on Activity Sheets 3.1A, 3.2A, 3.9C, and 3.9D are needed to complete this activity.

The following chart gives an interpretation to the meaning of dimension one, dimension two, and dimension three.

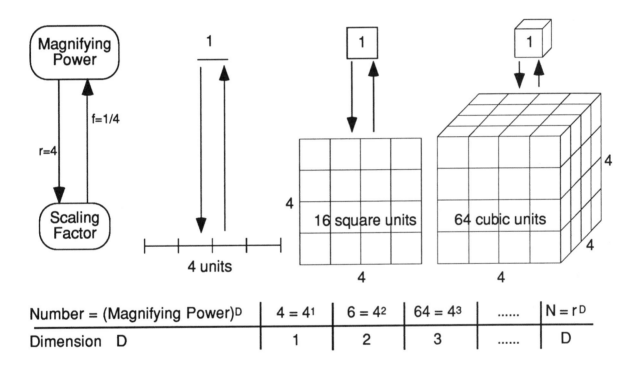

Number = (Magnifying Power)^D	$4 = 4^1$	$6 = 4^2$	$64 = 4^3$	......	$N = r^D$
Dimension D	1	2	3	......	D

1. Draw a series of three sketches similar to those above that represent the following numerical relationship.

$$3 = 3^1 \qquad 9 = 3^2 \qquad 27 = 3^3$$

What dimension is implied by each of your three sketches?

3.9B

On Activity Sheet 3.1A, each segment was replaced by N = 4 shorter segments configured in this pattern.

While the construction process can be continued *ad infinitum* , any particular segment at a given stage of the construction is r = 3 times longer than the four segments that will be drawn in its place at the next stage. The scaling ratio r represents the magnification power.

2. Study the constructions on Activity Sheets 3.1A, 3.2A, 3.9C, and 3.9D. Then find the number N of replacement segments and the scaling ratio r.

	Activity Sheet	Curve	N	r
a.	3.1A	Snowflake	4	3
b.	3.2A	Construction A		
c.	3.2A	Construction B		
d.	3.9C	Peano curve		
e.	3.9D	Zig-Zag		

> The similarity dimension of a self-similar fractal shape is given by the ratio of the log of the replacement ratio to the log of the scaling ratio.
>
> $$\text{Similarity Dimension} = \frac{\text{Log } N}{\text{Log } r}$$

3. Compute the similarity dimension for each of the five fractals constructed earlier.

      a. _____   b. _____   c. _____   d. _____   e. _____

4. Compare the similarity dimension for the first three fractals to those values computed by box counting in Activity 3.8 for the same three curves.

PEANO CURVE

The iterative construction processes on Sheets 3.9C and 3.9D generate two more self-similar fractals through scaling and substitution.

Replace each line segment with the pattern shown. Carry the construction through stage 3. The figure produced by iterating this construction process ad infinitum is called Peano curve.

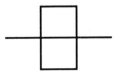

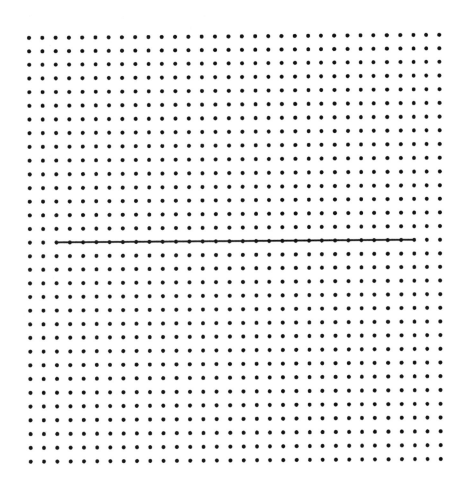

ZIG-ZAG

Replace each line segment with the pattern shown.
Carry the construction through stage 3.

Answers

UNIT 1   Self-Similarity

ACTIVITY 1.1A

Sierpinski triangle
Stage 4

1. At each new stage, more but smaller triangles are formed. In the limit, they are reduced in size to points.
2. A single, small triangle in the center, oriented as the original but 1/16 the size.

ACTIVITY 1.1B

Sierpinski triangle variation
Stage 2

1. The number of triangles becomes large without bound, but they are reduced in size to points.

2. Nine small triangles, in three separated sets of three each, all oriented as the original triangle.

ACTIVITY 1.2A

1. 1, 3, 9, 27, 81   2. 243; 3
3. 3^n; The number of triangles increases without bound.
4. 3/4, 9/16, 27/64, 81/256   5. 243/1024; 3/4
6. $(3/4)^n$; The area approaches zero.

ACTIVITY 1.2B

1. 1, 6, 36, 216
2. 1296, 7776; 6; The number of triangles increases without bound.
3. The triangles are increasing more rapidly in the Sierpinski triangle variation.
4. 2/3, 4/9, 8/27
5. 16/81, 32/243; 2/3; approaches 0
6. The area is decreasing more rapidly in the Sierpinski triangle variation.

ACTIVITY 1.3A

Square gasket
Stage 3

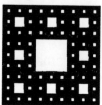

1. At each new stage, more but smaller squares are formed. The squares are reduced in size to points and the number of holes becomes large without bound. In the limit, none of the area of the original square remains.
2. 16 smaller squares, 4 in each corner

ACTIVITY 1.3B

1. 1, 8, 64, 512   2. 4096; 8
3. 8^n; The number of subsquares increases without bound.
4. 8/9, 64/81, 512/729   5. 4096/6561; 8/9
6. $(8/9)^n$; The shaded area approaches zero.
7. 1, 4, 16, 64   256; 4
   4^n; The number of subsquares increases without bound.
   4/9, 16/81, 64/729   256/6561
   $(4/9)^n$; The shaded area approaches zero.

ACTIVITY 1.4

Sierpinski tetrahedron
Stage 2

1. 64; 96; 34; 32
2. 64; 256; 4^n
   The number of tetrahedrons increases without bound while their volumes approach zero. Geometrically speaking, the tetrahedrons approach points.

3.

4.

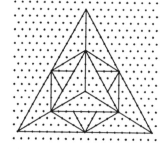

ACTIVITY 1.5

1.

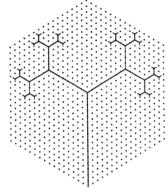

2. 4; 16; 1; 1
3. large without bound

4. yes

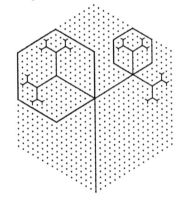

5. 2
6. 8
7. yes

ACTIVITY 1.6

1. The flowers, tightly clustered together in clumps on separate stalks, look very much like the clumps, tightly clustered together. This occurs repeatedly in larger and larger branches until the entire head is formed; yes; broccoli
2. large without bound
3. yes; yes
4. b; The cover is self-similar, but not strictly self-similar.
5. no; no
6. no; no; any part that does not contain leaves of the completed tree, for example the stem
7. yes; yes
8. yes; yes

ACTIVITY 1.7

1.   stage 2

stage 3

stage 4

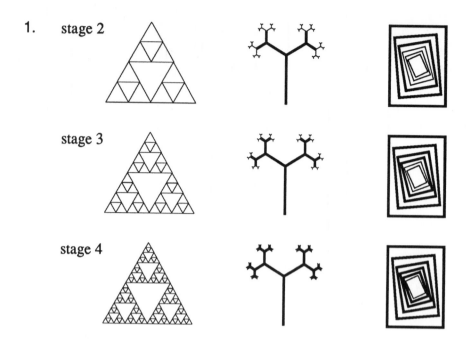

2.   in the center; at the ends of the branches; everywhere

ACTIVITY 1.8

1.   stage 1

stage 2

stage 3

stage 4

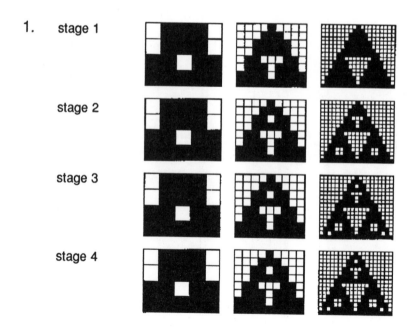

2.   stage 2; stage 3; stage 4
3.   stages 1 to 4: 12; 12; yes
4.   stage 1; It would take a higher stage for identical pixels to be lit.
5.   stage 1: 36; 38; no   stages 2 to 4: 36; 36; yes
4.   stage 2; It would take a higher stage for identical pixels to be lit.
7.   stage 1: 108; 124; no   stage 2: 102; 108; no
       stages 3 and 4: 102; 102; yes
8.   stage 3; It would take a higher stage for identical pixels to be lit.

ACTIVITY 1.9

1.   9; 10; 11; $n + 1$
2.   1   11   55   165 330 462 462 330 165 55   11   1
          1   12   66   220 495 792 924 792 495 220 66   12   1
3.   same numbers but in the reverse order
4.
5.    6.

7.   If the two cells above are colored the same, color the cell white.   If they are
       different, color the cell black.   End cells are always colored black.

ACTIVITY 1.10A

1.   yes; similar to stage 1 of the Sierpinski triangle
2.   The figure in the top four rows is replicated twice in the botton four rows;
       similar to stage 2 of the Sierpinski triangle
3.   Sierpinski triangle, stage 3   4.   32; 64
                                                             5. 2^n

ACTIVITY 1.10.B

1. (4,2); (3,4); (1,7)
3. black; white; black
5. (01100,10000); black

2. yes; white
4. (0111,1001); white
6. (011001,101000); white

ACTIVITY 1.11

1.
```
          1
         1 1
        1 2 1
       1 0 0 1
      1 1 0 1 1
     1 2 1 1 2 1
    1 0 0 2 0 0 1
   1 1 0 2 2 0 1 1
  1 2 1 2 1 2 1 2 1
```

2. similar to the stage-1 Sierpinski triangle variation

3.

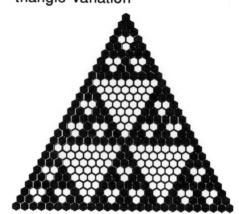

4.

ACTIVITY 1.12

1. If the two cells above it are colored the same, color the cell white.
   If the two cells above it are not colored the same, color it black.

2.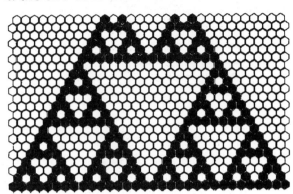

3. It is the same as the coloring pattern for Pascal's triangle starting at row 10.
   It is similar to stage 3 of the Sierpinski triangle.

4.   5.

6. 16; 16
7. similar to Sierpinski triangle stages 2 and 3

UNIT 2   The Chaos Game

ACTIVITY 2.1

1. triangle LTR
2. For a small number of points, they will appear to be randomly located. However, a pattern will emerge as the number of plays increases.

ACTIVITY 2.2

1. As the number of traces increases, they will begin to appear to cover the entire triangle LTR.
2. As the number of points increases, the Sierpinski triangle will begin to emerge.
3. yes

ACTIVITY 2.3A

1.    2. 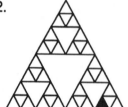   3.

4. RT   5. LTL   6. TRLT
7. the top of the top of the top of the top triangle
8. the left of the top of the left of the right triangle
9. the top of the right of the right of the left triangle
10. the left of the left of the left of the top triangle

ACTIVITY 2.3B

1. RT   2. LLR   3. TLR

4.   5.   6.

7. LRR;   TLL;   TTT;   TRL: RTL
8. LLLL;   LRTL;   TLRT;   TTLL;   RLTT;   RTLL

ACTIVITY 2.4

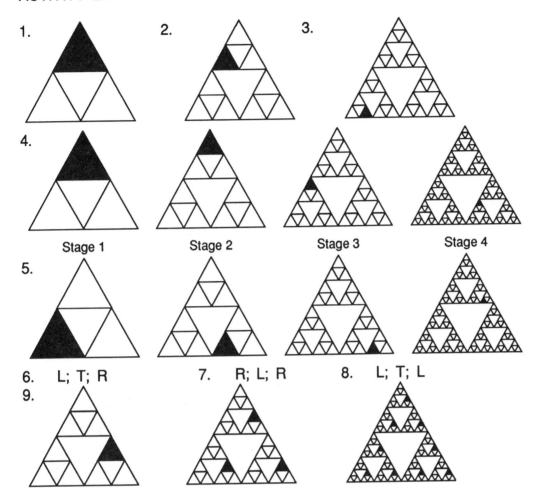

Stage 1   Stage 2   Stage 3   Stage 4

6. L; T; R   7. R; L; R   8. L; T; L

9.

ACTIVITY 2.5

1. LT, TT, RT, LR, TR, RR
2. 27; LTT, TTT, RTT, LRT, TRT, RRT
3. 81; 27; 9; 3
4. LRTT, TLRT, TTLR, RTTL, RRTT, RRRT, LRRR, LLRR, TLLR, RTLL, LRTL, TLRT, RTLR, LRTL, RLRT, RRLR, TRRL; 15
5.    6. yes; 3^n; yes

ACTIVITY 2.6

Answers will depend on individual experimental results.

ACTIVITY 2.7

1. 1/9   2. 1/27   3. 1/81
4. 1/3; 1/9; 1/27
5. 1/9   6. 1/27   7. 1/81
8. probability approaches 1; probability approaches 1
9. 1/18   10. 1/54   11. 1/216
12. 1/24; LRR; RLR
13.    14. 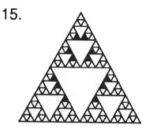   15.

16. LRRR, RLRR, RRLR, RRRL
17. RRRRR; 1/32
18. LLLLLL; 1/46,656
19. The Sierpinski triangle will emerge, but with an uneven distribution of points.
20. same as 19.

ACTIVITY 2.8

1.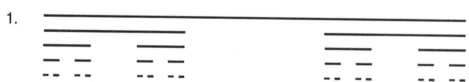

2. 32; 2^n
5. The tree is self-similar. The set of leaves on the tree is strictly self-similar.
6. $LLL\bar{L}...$, $LRR\bar{R}...$; $RLL\bar{L}...$, $RRR\bar{R}...$
7. $LLL\bar{L}...$, $LLR\bar{R}...$; $LRL\bar{L}...$, $LRR\bar{R}...$; $RLL\bar{L}...$, $RLR\bar{R}...$; $RRL\bar{L}...$, $RRR\bar{R}...$
8. yes
10. 0.11   11.   0.001   12.   0.0111
13. The cardinality of the Cantor set is the same as that of the interval [0,1].

ACTIVITY 2.9

1. 27; 81; 3^n
2. The number of subintervals increases without bound while their width decreases
    to zero.

3. The tree is self-similar. The set of leaves on the tree is strictly self-similar.
4. LLLL̄... , LRRR̄... ; TLLL̄... , TRRR̄... ; RLLL̄... , RRRR̄...
5. yes
6.

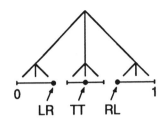

7.

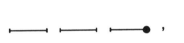 , TRRR̄... ,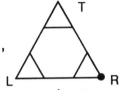

8.

 , RRRR̄... ,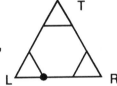

9.

, LRRR̄... ,

11. LTT, TLR, RTL
12. stage 2: $3 + 9/2 = 15/2$;
    stage 3: $3 + 9/2 + 27/4 = 57/4$;
    stage n: $3 + 9/2 + 27/4 + ... + 3 (3/2)^n = 3 (1 + 3/2 + ... + (3/2)^n) = 9 (3/2)^n - 6$
13. no

UNIT 3   Complexity

ACTIVITY 3.1A

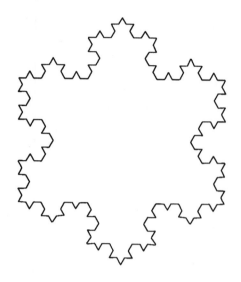

ACTIVITY 3.1B

1.

Stage	0	1	2	3	4	5	...	n
Segments	3	12	48	192	768	3072	...	$3 \cdot 4^n$

3.

Stage	0	1	2	3	4	5	...	n
Perimeter	3	4	16/3	64/9	256/27	1024/81	...	$3 \cdot (4/3)^n$

4. ∞

5.

Stage	Area
0	$\dfrac{\sqrt{3}}{4}$
1	$\dfrac{\sqrt{3}}{4} + \dfrac{\sqrt{3}}{4} \cdot \dfrac{3}{9}$
2	$\dfrac{\sqrt{3}}{4} + \dfrac{\sqrt{3}}{4}\left(\dfrac{3}{9} + \dfrac{3 \cdot 4}{9^2}\right)$
3	$\dfrac{\sqrt{3}}{4} + \dfrac{\sqrt{3}}{4}\left(\dfrac{3}{9} + \dfrac{3 \cdot 4}{9^2} + \dfrac{3 \cdot 4^2}{9^3}\right)$
4	$\dfrac{\sqrt{3}}{4} + \dfrac{\sqrt{3}}{4}\left(\dfrac{3}{9} + \dfrac{3 \cdot 4}{9^2} + \dfrac{3 \cdot 4^2}{9^3} + \dfrac{3 \cdot 4^3}{9^4}\right)$
5	$\dfrac{\sqrt{3}}{4} + \dfrac{\sqrt{3}}{4}\left(\dfrac{3}{9} + \dfrac{3 \cdot 4}{9^2} + \dfrac{3 \cdot 4^2}{9^3} + \dfrac{3 \cdot 4^3}{9^4} + \dfrac{3 \cdot 4^4}{9^5}\right)$

6. a.
$$\dfrac{\sqrt{3}}{4} + \dfrac{\sqrt{3}}{4}\left(\dfrac{3}{9} + \dfrac{3 \cdot 4}{9^2} + \dfrac{3 \cdot 4^2}{9^3} + \dfrac{3 \cdot 4^3}{9^4} + \dfrac{3 \cdot 4^4}{9^5} + \ldots + \dfrac{3 \cdot 4^{n-1}}{9^n}\right)$$

b.
$$\dfrac{\sqrt{3}}{4} + \dfrac{\sqrt{3}}{4}\left(\dfrac{3}{9} + \dfrac{3 \cdot 4}{9^2} + \dfrac{3 \cdot 4^2}{9^3} + \dfrac{3 \cdot 4^3}{9^4} + \ldots\right) =$$

$$\dfrac{\sqrt{3}}{4} + \dfrac{\sqrt{3}}{4}\left(\dfrac{\frac{1}{3}}{1 - \frac{4}{9}}\right) = \dfrac{\sqrt{3}}{4} + \dfrac{\sqrt{3}}{4}\left(\dfrac{3}{5}\right) = \dfrac{2\sqrt{3}}{5}$$

7. Infinite perimeter and finite area

8. The Koch curve is both self-similar and strictly self-similar. But - the Koch snowflake is not self-similar. There is no section of this curve which looks like the whole one.

ACTIVITY 3.2

1.   Construction A   Construction B

2.

Stage	Segment Length	
	Constr. A	Constr. B
0	1	1
1	1/3	1/4
2	1/9	1/16
3	1/27	

3.   Scaling factor   Construction A: 1/3
                          Construction B: 1/4
4.   1/3
5.   Curve B appears to be more complex.
6.   Curve B

ACTIVITY 3.3

1. 2.   Standard   Semilogarithmic   Double Logarithmic

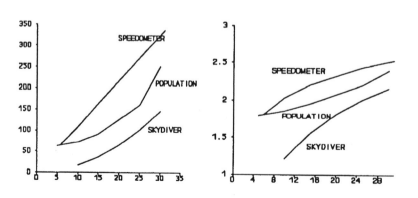   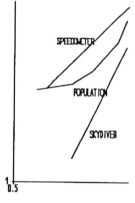

3. The "speedometer" data appears to lie on a straight line when it is plotted on standard graph paper suggesting that the time/distance relationship is a linear relationship.
4. The "population" data appears to lie on a straight line when it is plotted on semilogarithmic graph paper suggesting that the time/population relationship is exponential.
5. The "skydiver" data appears to lie on a straight line when it is plotted on double logarithmic graph paper suggesting that the time/distance relationship is a power relationship.
6. Slope = 2

ACTIVITY3.4

1.

Linear	Exponential	Power

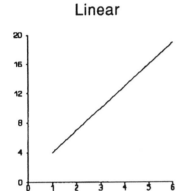 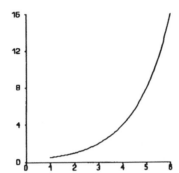 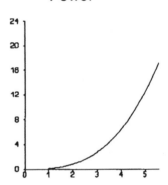

2.

x	1	2	3	4	5	6
y	.5	1	2	4	8	16
log y	-.301	0	.301	.602	.903	1.204

3.

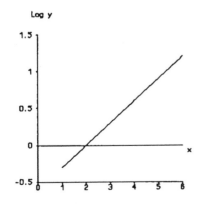

4.

x	1	2	3	4	5	6
y	.01	.08	.27	.64	1.25	2.16
log x	0	.301	.477	.602	.699	.778
log y	-2	-1.097	-.569	-.194	.097	.334

5.

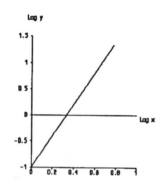

6. $y = k\,a^x$

  a. $\log y = \log k\,a^x$

            $= \log a^x + \log k$

    $\log y = x \cdot \log a + \log k$

  b. $Y = A\,X + K$

  c. Slope A equals the log of the base a in the original exponential function, while the y-intercept K equals the log of the original coefficient k. Accordingly, $a = 10^A$ and $k = 10^K$.

7. $y = k\,x^r$

  a. $\log y = \log k\,x^r$

            $= \log x^r + \log k$

    $\log y = r \cdot \log x + \log k$

  b. $Y = r\,X + K$

  c. The slope r equals the exponent in the original power function, while the y-intercept K equals the log of the coefficient k in the power formula. Accordingly, $k = 10^K$.

8. a. No. However, the base a and coefficient k would be computed by $a = e^A$ and $k = e^K$.

  b. No. However, the coefficient k would be computed by $k = e^K$.

ACTIVITY 3.5B

2. The data is best fit by a linear function. Equivalently, a power function with exponent equal to 1 gives essentially identical results since such a power function would in fact be linear.
3. The data is best fit by the exponential function $y = 42.493 (1.057)^x$

ACTIVITY 3.6

For all boxcounts only good approximations are required - not exact numbers!

1. WAVE

1 / scale x	4	8	12	16	24	32
Boxcount y	11	34	51	68	100	135

BLACK HOLE

1 / scale x	4	8	12	16	24	32
Boxcount y	16	60	120	208	448	756

FUNCTION

1 / scale x	4	8	12	16	24	32
Boxcount y	9	24	45	63	105	150

2.

WAVE	BLACK HOLE	FUNCTION

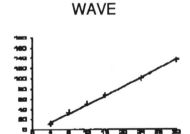 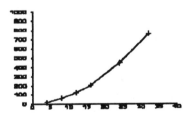 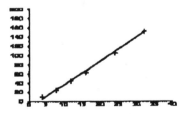

3.

| WAVE | BLACK HOLE | FUNCTION |

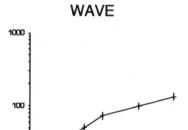
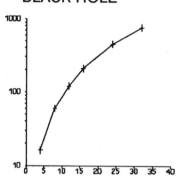
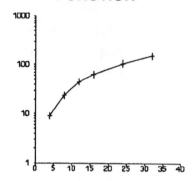

4.

| WAVE | BLACK HOLE | FUNCTION |

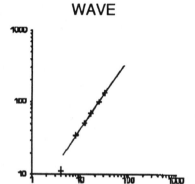
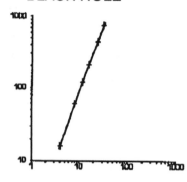
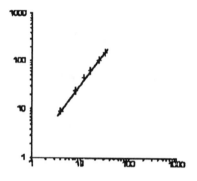

5. Power relationship (for Wave also a linear relationship exists)

7.

	WAVE	BLACK HOLE	FUNCTION
Box dimension	≈ 1	≈ 1.85	≈ 1.35

ACTIVITY 3.7A

1.

1 / scale x	4	8	12	16	24	32
Boxcount y	16	48	84	122	194	283

2.

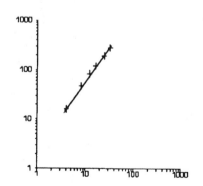

3.   Box dimension   ≈1.37

ACTIVITY 3.8A

1.   SNOWFLAKE

1 / scale x	4	8	12	16	24	32
Boxcount y	6	14	26	32	55	74

3/2 CURVE

1 / scale x	4	8	12	16	24	32
Boxcount y	10	26	41	63	104	155

COMPOSITION

1 / scale x	4	8	12	16	24	32
Boxcount y	15	43	68	103	188	255

2.   SNOWFLAKE   3/2 CURVE   COMPOSITION

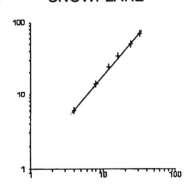 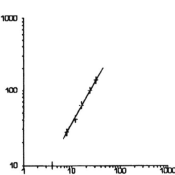 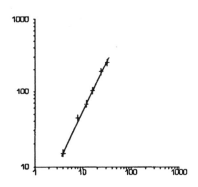

3.

	SNOWFLAKE	3/2 CURVE	COMPOSITION
Box dimension	≈ 1.2	≈ 1.29	≈ 1.28

4.   The section having the greatest complexity determines the box dimension for the whole figure.

ACTIVITY 3.9B

2.

	Activity Sheet	Curve	N	r
a.	3.1A	Snowflake	4	3
b.	3.2A	Constr. A	5	3
c.	3.2A	Constr. B	8	4
d.	3.9C	Peano curve	9	3
e.	3.9D	Zig-Zig	6	4

3. a.   1.26
    b.   1.46
    c.   1.5
    d.   2
    e.   1.29

ACTIVITY 3.9C

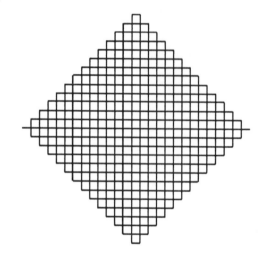

ACTIVITY 3.9D

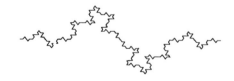

Fractals for the Classroom
Strategic Activities Volume 1
Slide Set

1. Broccoli Romanesco
A new breed of broccoli exhibits the feature of self-similarity to a surprising extent. Observe how all its roses look like small copies of the whole broccoli.

2. Sierpinski Snail-shell
Very recent biological models have tried to establish these shell patterns by cellular automata. Indeed a shell like this grows layer by layer. Image granted by Peter Plath, Universität Bremen.

3. The Chaos Game
Three stages of the chaos game which generates the Sierpinski triangle on a PC: (right) stage at 1500 iterations, (left) stage at 4500 iterations, (top) final stage.

4. Pascal's Triangle
A color coding experiment for Pascal's triangle. The familiar number pattern is encoded into the coloring of a pyramid of hexagons which represent the entries c of the triangle. The coloring is

     dark blue   —if $c \bmod 2$ is equal to 1 (i.e., c is odd),
     light grey   —if $c \bmod 4$ is equal to 0,
     light blue   —if $c \bmod 4$ is equal to 2.

5. Sierpinski Tetrahedron
A raytraced rendering of a 3D version of the construction similar to the one generating the Sierpinski triangle. This is a strictly self-similar object: It is composed from four parts, each of which is half the size of the original. Image courtesy of Daryl Hepting, Alan Snider, and Przemyslaw Prusinkiewicz, University of Regina, Regina, Canada.

6. The Mandelbrot Set 3D
A 3D rendering of the Mandelbrot set which is based on a distance-estimator algorithm (i.e., contour lines of the surface correspond to lines of the same distance to the Mandelbrot set). This image is computed to an ultra-high resolution of 12 million pixels.

7. Secco (Mandelbrot Set)
Although the Mandelbrot set is not self-similar, it contains infinitely many small copies of itself. This close-up shows one of the most prominent ones which can be found close to the imaginary unit $i = \sqrt{(-1)}$ in the complex plane. Colors in the image encode distance to the Mandelbrot set.

8. Escalante 3D
A close-up of the 3D rendering of the potential of the Mandelbrot set. This is a high-resolution still picture from the video *Fractals: An Animated Discussion* (by H.-O. Peitgen, H. Jürgens, D. Saupe, and C. Zahlten, Freeman, New York, 1990) which shows this image in a spectacular flight animation.

9. Yellowstone Lake 3D
The coloring of this 3D rendering of the potential of the Mandelbrot set was inspired by winter scenes at Yellowstone Lake. The clouds in the background are created by a random fractal algorithm.